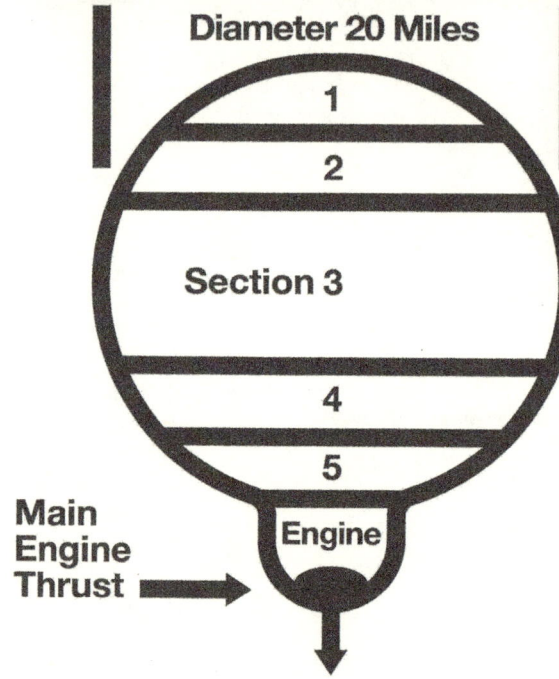

MIGRATION 2470

Earth's **Dystopian** Future
A Sad Ending
An Even Sadder New Beginning

written by

John (chief) Adams

Migration 2470

Copyright © 2023 John B Adams

All rights reserved. No part of this publication may be reproduced, distributed, or transmitted in any form or by any means, including photocopying, recording, or other electronic or mechanical methods, without permission of the publisher, except in the case of brief quotations embodied in critical reviews and certain other noncommercial uses permitted by copyright law.

Any reference to historical events, real people, or real places are used fictitiously. Names, characters, and places are products of the author's imagination.

Contents

Introduction		vii
Chapter 1	Pre-Launch Countdown	1
Chapter 2	Fuel raid	9
Chapter 3	The Launch	11
Chapter 4	Earth's pursuit	15
Chapter 5	The Early Years	25
Chapter 6	The Balkanization of the crew	37
Chapter 7	Slow Times – the Middle Voyage	47
Chapter 8	Sam's Cloning	49
Chapter 9	Mystery of the Middle of Their World	55
Chapter 10	Thrust Reverse	61
Chapter 11	Sam's Dilemma	67
Chapter 12	New Alliance	73
Chapter 13	The Sergeant	83
Chapter 14	Little Known Areas	87
Chapter 15	The dark web and the ship flip	95
Chapter 16	Sam's New Duties	107
Chapter 17	Earth's surprise	111
Chapter 18	Sam Makes His Move	115

Chapter 19	Conspirators	121
Chapter 20	The Setup	125
Chapter 21	Mutiny Planning	129
Chapter 22	Speeding Up the Takeover	139
Chapter 23	Changes	145
Chapter 24	New Faith	155
Chapter 25	End Game	161
Chapter 26	A Big Surprise	167

Introduction

In the year 2150, Earth reached its inevitable crossroads. Conditions had deteriorated in every corner of the world. Communication, manufacturing, and trade worldwide were coming to a grinding halt. The world's population had grown so large efforts to conduct a worldwide census had failed. All the world's governments (and now there were many) had not been able to provide stable conditions nor reverse climate change, either of which would have allowed their citizens to prosper.

There were countless armed conflicts in progress over the Earth's resources, and new ones starting every day. Environmental conditions had deteriorated to the point that crop yields did not produce enough food to sustain the world's population. Eventually, the world's elite academic liberal ruling class began to feel the effects, and they all had the same epiphany. A little cooperation and common sense would go a long way toward solving some of the problems facing them all.

Their solution was to eliminate anyone or anything that stood in their way. Consolidation by any means justified the end results. That became their rallying cry. So began an exceptionally long and bloody process towards a one-government world. The world's leaders had managed to create the most oppressive and murderous government in human history.

Organized religion was soon stamped out, and this had a devastating effect on society. This released the savagery that lurks in all of us when one's moral compass is missing. But there was a second chance looming – it was called the Project.

Character key: *Migration 2470*

(A character's rank, and position may change, but not their names)

Chief of Operations: military: Samuel "Sam" Otto Bernard, Chief S.O.B.

Ship Captain: military: Dmitri Volkov ("The wolf")

Ship Security: military: Colonel John Hooker

Commander, Computers and Cloning research: Thomas "Tom" Hacker

Ships, Chief Engineer: military: Wolfgang Schmitz

First Mate: military: Oliver Whipple

Medicine and Science officer: Dr. Emily Harris

Head of Maintenance: civilian: Qiao Hui

Food, flora, and fauna: civilian: Anastasia

Board Chairman: civilian, politician: Nathan Brock

Sergeant: military: Toby Sawyer (wife, Janice; daughter, Alice; son, David)

BC: bridge committee: The Captain, Sam, Whipple, Chairman Brock, Colonel Hooker, Major Sawyer and Thomas (Tom) Hacker

Migration 2470

CHAPTER 1

Pre-Launch Countdown

The Chief of Operations sat at his desk as he had done almost every day for the last thirty-plus years.

He never knew the people who were in this very same office before him. He had been recruited so many years ago it seemed like a different place and time, and in many ways, it was. He was having doubts about the Project and the direction it had taken. He and many others hoped the Project would provide humanity with a second chance. The best he could tell, the Project started 60 years ago, around the year 2410. He did not know exactly how many years the Project had been in existence before he had been recruited. More time had passed before he was recruited than was left before completion.

The intent had been to save humanity from itself by finding another planet and making it available for colonization. He now sat looking out of the window of his office inside the largest building that would ever be built on planet Earth. Ninety-nine percent of the Earth's population, estimated at fourteen billion people, did not know of the Project's existence, much less its purpose.

At the beginning of the project, he had been led to believe it had started innocently enough. Earth's population had increased to a point where all signs of human decency had disappeared. Across the human spectrum, people were trying to destroy each other over the Earth's shrinking resources. Governments were becoming day by day less relevant and were being swallowed up by the world's business corporations who were becoming larger and larger as they merged. They soon controlled all the world's industries and resources. When they reached the point that only one corporation remained, the Project was born. It was the name that stuck.

The defense industry wound up being the last functioning corporation and had control of all the industries and natural resources in the world. The globalization of Earth has finally been achieved, and the last corporation did not have any problems gaining control and keeping it. They also manufactured and controlled all the weapons in the world, including weapons of mass destruction, and did not hesitate to use them on anyone who might threaten them.

Chief of Operations Samuel O. Barnard became known as the Chief S.O.B. This was a well-deserved nickname. The Project, in the early days, started as a plan to send only a small, unmanned probe, and the knowledge gained would eventually help all of Earth and would not have to be kept a secret. This would eliminate problems concerning security. Sam and the Corporation soon realized that without a faster-than-light drive for the ship, it would take too long for a round trip. The nearest three gold-D-locks planets, which showed the most promise, were approximately 160 light years away from Earth, and two of them orbited a sun, designated K2-239, that was comparable to Earth's Sun. Both planets had approximately the same mass as Earth. The third planet, also the size of Earth, orbited a red dwarf, designated K2-240.

The corporation engaged in numerous discussions regarding its next steps. Sam's predecessor was thought to have been behind the idea of sending a probe, but then, because of time constraints, switched to a one-way ship. The vast difference between building a ship big enough to house thousands of people and building a probe would have to be kept secret at all costs.

A large, self-sustaining ship would allow the expedition to explore many new worlds and alter the final destination if needed. The expedition's personnel all realized they would have to come to grips with the fact that when they left Earth, it was forever. They would never see or hear again about the Earth and its people. The more Sam thought about the Project, the more he realized that he felt like he was deserting his fellow man. The thought embarrassed him, but this was a feeling that was new to him and would have to be ignored.

Sam has spent almost all his efforts in the last 30 years keeping the Project a secret. When the competition for resources became increasingly fierce, it became obvious to Sam why the board of directors had chosen him to head up the Project. He had spent his early years in the spy business. Sam had been the last director of America's intelligence community. The various protocols he had put in place for the Project had cost the lives of thousands of workers and their families. The shortages created in the world's resources because of the Project's need to build an intergalactic ship, a sphere with a diameter of approximately twenty miles, was responsible for the death of millions more.

Leaks in the Project's security and mounting insurrections around shuttle sites were becoming increasingly more of a concern. As the true scope and size of the Project leaked out it became obvious to all employees, even the lowest levels, that many people were going to be left behind when the Project came to fruition.

The alarm woke Sam up on the couch in his office. He went over to the console, where the surveillance cameras were located and took a seat. The recent attempts to break in were caused by all the rumors of what was about to happen in five days. Sam was as anxious as anyone to leave Earth. Because of years of secrets and compartmentalizing, he was the one person who actually had a clear view of Project "Migration." Many problems had not been solved, the most pressing being what he was watching unfold on the security monitors. Some of his security personnel appeared to be in a firefight with external entities. Sam had authorized the use of deadly force. This latest incursion would not last long, nor would it be successful.

The most important problem that would have to be solved for an estimated 600-year mission to succeed was time dilation. The unknown factor was that no one actually knew how many years would pass on the ship versus time passage on Earth. The success of the migration depended on being able to clone people, livestock, and plants – thus, the voyage's length became less important.

The physical part had been perfected for years, but the procedures necessary to capture and store a conscious human mind and then reintroduce it into a new body were still in its infancy. There had not been time to do any long-term trials, such as more than three or four generations, performed in around 30 years.

The loading of everyday provisions (food, water, and raw materials) was almost on schedule. All the machinery needed to manufacture repair parts for the ship's systems was not yet on board. The ship's crew was going to be a cross-section of people who had the knowledge to keep the ship operating during a long voyage.

Sam had run across a problem that neither he nor anybody else had anticipated. A sizable number of workers who were discreetly recruited as crew members were reluctant to depart from Earth. *New Hope* had been under construction for more than seventy-five years. The ship offered an improved standard of living that was accessible exclusively to those employed by the Project. This was a shock, but everyone working in crew recruitment quickly came to the same conclusion: the workers and their families needed to disappear to maintain secrecy. If the unwilling recruit came from a particularly small pool of qualified workers who were hard to replace, they were kidnapped along with their family and placed on board the ship and their consciousness was put in storage. The workers would then be available for use if needed during the voyage.

Sam almost single-handedly had been working on the details of how the ship should function. He decided that he would not be the captain and would not change his title as Chief of Operations. The man presently in charge of the ship, Dmitri Volkov, would be named the permanent captain. Dmitri had been the last commanding general of the Russian armed forces. He joined the crew 45 years ago, a few years after the ship became habitable in space.

It was not known outside of a small circle of people that Dmitri had recently undergone cloning on board *New Hope*, and the procedure had been successful. With this added information, the cloning issue was put to bed and would not by itself hold up the launch. Everyone was under the impression it was his first cloning, but Sam might be the only one alive who knew the actual number. The captain had been cloned three times previously.

Sam did not think that it was necessary to launch in five days. He had a staff prelaunch meeting later this afternoon. He had not had anything to eat since waking up this morning in his office. He ordered a late lunch and sent out a request to all department heads to be in his office at 1,800 hours or be ready to meet remotely.

He went over the list of department heads: Captain Dmitri Volkov, Chief Engineer Wolfgang Schmitz, First Mate Oliver Whipple, Head of Security Colonel John Hooker, Medical Officer Dr. Emily

Harris, Head of Maintenance Qiao Hui and Anastasia Kovalenko, Food Production.

There were issues involving the ship's personnel and a senior officer. Anastasia was a recent addition. She had been hired to replace the last head of food production, who eventually realized that he would not return to Earth. He died in a shuttle accident returning to Earth. His shuttle just disappeared from the radar screens.

The meeting began.

"When the last of my fuel supply arrived, I would need two days, and then the ship would be ready," announced the Captain.

"Yes, I will also be ready," Wolfgang agreed.

"I will also be ready," Oliver said.

Hooker's report was not well received.

"I'm having enormous problems with the workers who were still on board. As the preparations for departure are completed, the number of workers we need grows smaller. The question of what to do with the worker surplus is becoming an issue. I can't just send them back to Earth. There is nothing to keep them from telling the world what their governments and the large corporations have done to the resources of planet Earth!" he said.

"John, offer them the chance to be ready for cloning," instructed Sam. "If we need them during the voyage, they will be awakened. Eliminate those who resisted."

The Chief S.O.B. had no plans to waste any resources on the lowly workers.

"I will gladly accept the workers' bodies," offered Dr. Harris. "This would add greatly to the ship's store of human genetic material."

Anastasia nodded. "I'm in full agreement with you there."

Qiao Hui was the most cautious. He was in charge of manufacturing the necessary fuel. The plan was to use a breeder reactor because it would produce more radioactive fuel than it used, thus allowing the use of a fusion-powered rail gun-type propulsion unit.

With the rail gun, you fire enough electrons in a condensed stream, and you will get an opposite reaction and have useable thrust. This was especially true for a ship operating in the vacuum of interplanetary space. There is absolutely no drag present. The ship would accelerate for the first half of the trip. At mid-point, they would flip the ship around and restart the engine. This would decelerate the ship at the same rate as it accelerated the first half of the voyage and bring the ship to the new planet.

CHAPTER 2

Fuel raid

Lately, the fuel arrivals have not been keeping up with the construction of the fuel storage areas. This would have to be addressed before leaving orbit. Uranium fuel rods would be needed to power the shuttles that were going to be taken along on the voyage.

Sam had developed quite a simple plan concerning the fuel issue. It was to do nothing until the fuel storage areas had been completed. He summoned Colonel John Hooker to enlighten him about his role in the fuel plan.

"John, the fuel storage areas are ready. Take sufficient personnel and equipment to Earth, with the objective of taking control of one of the two remaining operational nuclear plants. Confiscate the available fuel at the plant. And... use as much force as necessary," Sam said.

***

The raiders came in three attack shuttle crafts, with three hundred men in each craft. Colonel Hooker expected stiff resistance from the fuel plants' security forces as they had grown quite large. He was not disappointed – several hundred security guards were killed. The Colonel's men also suffered heavy losses, 319 casualties, to be exact.

But beforehand, Sam and John had a private conversation, and they had agreed the security force needed to be reduced by around 300 men.

"I will plant a bomb on one of the shuttles. If I lose less than 300 soldiers, I'll detonate it," John said.

They both chuckled. "Trying to kill two birds with one bomb, hey," commented Sam.

The two cargo fuel carriers returned unscathed; Chief Engineer Qiao Hui was happy.

***

It had been 22 days since the meeting; the fuel raid was 18 days ago. The fuel theft had galvanized what was left of Earth's military. Earth was in the process of trying to put together a strike force and go after the raiders. Hundreds of shuttles left on Earth that could be used to reach *New Hope*. The ship was parked on the dark side of the moon. The Project had realized that such a thing might happen.

Project soldiers who had been wounded during the fuel raid and left behind had, in return for medical treatment, shared all the information they processed about the ship *New Hope* and the plan to leave.

CHAPTER 3

The Launch

The board that headed up the project was now on board *New Hope*. Some of the board members were relatively recent additions, while others had tenures extending up to 25 years.

New Hope was parked on the dark side of the moon. The ship's engines started on January 1st, 2470. The acceleration rate during the initial blastoff was going to be terribly slow; it seemed like hours passed before the crew even felt the ship move at all. Nothing about the launch required the crew, except the engine crew, to make any preparations for the launch. It was days before everyone seemed to accept the fact that they were actually now on their way.

The shuttles left behind on Earth could easily overtake *New Hope* unless the ship had at least an estimated three-month head start. The board and Sam also realized this possibility, so precautions had been taken. Any pursuing shuttles could be scuttled from the bridge of *New Hope* if they got within 40,000 miles.

The plan was quite ingenious. No explosives were involved; a virus had been built into the circuitry when the shuttles were manufactured. On command, the circuitry controlling the cooling system would raise the upper engine temperature to an unsustainable high. After the signal was received, the shuttles' engines would explode in a matter of a few seconds.

***

Sam woke up in his quarters on *New Hope*. Today was going to be a fresh start for humankind. The voyage had started; the main engines were now online. From the ship's present location on the dark side of the moon, it was going to take about five years if only the main engines were used to pass Pluto, our outermost planet. This would be the scenario if there were no pursuit from Earth's forces.

However, if any should arise, the board has made provisions to shorten the time of travel. Shuttles were attached to the outside of the ship for the trip, and they were all packed full of supplies. The design engineers and Colonel Hooker decided to position them in such a way as to provide propulsion in any direction. This extra push was predicted to cut the ship's escape time from the solar system in half. Most importantly, the shuttles provided maneuvering thrust should it be necessary. The biggest drawback was that one-sixth of the shuttles would be out of fuel.

The corporation decided to start the clock anew when the engines fired. Year One would then begin. Sam was on the bridge along with the captain and the chairman of the board.

This is the first time the three of us have been physically together, Sam thought.

All the previous meetings were via a video setup. Sam knew he served at the pleasure of the board. The subject he was about to bring up was going to be touchy.

"Captain, won't you and Chairman Brock please join me in the Captain's ready room adjoining the bridge?" Sam asked.

When the three of them sat down, the Captain looked at him expectantly. "Well, Sam, we are waiting," he said.

"Certainly, Captain. At any one time, two of us need to be awake to protect the third. The ship requires a minimum crew of 2,100 people to be awake all the time. The cloning storage facility holds approximately 75,000 souls. The plan, before launch, called for the crew, when they reached the end of their natural life, to be made ready for cloning and placed in storage until their turn to return came up. But this is the problem. We can accomplish cloning the physical body after a natural death, sickness, or an accident. But, to store someone's mind for the future, the process had to start before the body died. Everyone awake on board the ship had, for the most part, known about this problem. Unfortunately, the crew was nearly all type-A personalities. They would undoubtedly be skeptical about giving up their lives with only a promise of being cloned at a later date."

So, they made a new plan. The ship would be divided into sections. The Captain would be in charge of all the ship's operations, and Chairman Brock would be in charge of everything else. The two halves would allow the ship's crew to be separated, and day-to-day contact would be limited as much as possible. Access from one half to another would be restricted by several controlled points that could be managed if necessary.

"Well, gentlemen, I suggest we part ways for now so that each could go and prepare for the long journey," said Chairman Brock.

Sam was not ready to trust Brock – the man remained an enigma. Brock had given off bad vibes during the meeting. He suspected the Captain may have come away feeling the same way.

When you met people for the first few times who had been cloned multiple times, it was impossible to learn very much about their history unless they wanted you to know. His record was sparse;

Sam had not seen any information that would instill confidence. Sam remembered reading a report on his background prepared in secret by a tech. Brock had been involved in all kinds of semi-illegal ventures throughout the years.

Brock is another department head about whom there was extraordinarily little information available that would justify putting him in charge of the Migration mission, Sam thought.

CHAPTER 4

Earth's pursuit

The captain was especially anxious about any pursuit from Earth now that the ship had launched. Earth would soon be able to see the ship. If the extra boost from the shuttles was going to be needed, it was going to take time to bring them online. This had not been practiced before, so neither Captain Volkov nor the First Mate Oliver Whipple knew how much time would be needed.

How long was it going to take to get the shuttles online and up to speed? *New Hope's* engines had been online for 20 hours. The ship had just reached a 0.3g targeted acceleration rate. This was the speed the chief engineer needed to be able to perform the main engine testing. *New Hope* was now at a point that Earth was now in full view. Volkov was astonished at what he saw and heard on the radio. Earth must have known about what the Corporation was about to do.

Volkov called the chief engineer. "Increase the rate of acceleration to 1G, equating to an increase of 22mph per second," he ordered.

"But we haven't put the particle beam engine through all the tests yet! I will need another 10 hours," he complained. "So, my engine performance predictions were educated guesses."

Nevertheless, he started the acceleration as instructed.

Volkov called up the Colonel, who had been tracking the ships on the radar. "Captain, if the ship maintained its present rate of acceleration, would the Earth's ships catch up to us? And, if so, would they be inside of 40,000 miles?"

Sam, after hearing the Captain's request, started to counter the order but realized they had no choice. At 1G of acceleration, *New Hope* would approach the speed of light in about one year. They would be plowing new ground.

The theory was that constant acceleration drives could be used to approach relativistic speeds. From *New Hope's* planetary frame of reference, its speed will appear to be limited by the speed of light – it will approach the speed of light but never reach it. This means that for voyagers on *New Hope*, the journey will appear to be much shorter than what planetary observers see. Relativity was going to come into play, and no one knew what to expect. Whatever was going to happen when they approached light speed was something that still needed to be researched. Hopefully, plans could be made to deal with it when it occurred.

Sam did not know whether he should be worried or not.

At this point, Volkov decided to call Sam and the Chairman. Both men came to the bridge. The three men were not expecting the Earth's response to be this fast or massive.

"I'm not 100 percent sure. My numbers suggest the closest any of the ships would get would be 35,000 miles," said the Colonel. "This prediction had a possible error of 5 to 8 percent on either side of 35,000."

The option of using one-sixth of the shuttles to increase the ship's acceleration was discussed, and it was not a viable option. The fuel issue had come up again. No action needed to be taken yet; they would just have to continue.

The ships in pursuit were burning up the airways, trying to get the migration ship, *New Hope*, to respond. Sam had been monitoring their transmissions. He and the others were trying to figure out how much Earth knew about the Project, the existence of the sleeping virus and the fact that *New Hope* had no offensive weapons. Since the shuttles were constructed over a period of 60 years, some of the earliest shuttles incorporated older technologies. The new virus used was not compatible. Most of the older shuttles had been decommissioned, but not all.

Before the virus could be activated, they were going to have to do a complete check on the 1,800 shuttles attached to *New Hope's* hull. If the virus were activated and a shuttle destroyed itself while attached, that would destroy *New Hope*.

"I had not realized a shuttle could be destroyed remotely even if its engine were off," said Sam.

Volkov instructed the Chief Engineer on how to search for the computer logs. The knowledge of the virus's installation had been on a need-to-know basis. The virus for years had been used to eliminate people who were a problem or a potential problem. This use of the virus for years would explain many mysterious shuttle explosions and disappearances.

The Chief Engineer started immediately. Fortunately, a code was used in each shuttle's serial number, making it easy to identify any shuttles with the virus. The search was going to take him about 96 hours.

The Colonel updated his numbers. He had had a chance to compare *New Hope's* acceleration rate to the Earth's ships, and a decision would have to be made in approximately 100 hrs.

Sam, the Captain, and the Colonel, over their first meal since leaving, began to discuss options. The present situation had not been expected and there were no plans in place. The thing they hoped would happen was that the pursuit ships would not be able to overtake them and, therefore, would not pose a threat.

The big question was what they were going to do if they did get close enough to be a problem. The easiest solution would be to use the virus, but they could not be sure if all the pursuit vessels had the virus installed.

"I have an idea," said Sam. "But I have to wait for Wolfgang to finish his search."

That is what he told the others. They finished their meal and agreed to meet when Wolfgang had finished his search. Sam found Colonel Hooker in his office and brought him up to date.

"Colonel, how many nuclear bombs are on board and ready for use?" Sam asked.

"I don't know, but I'll find out and let you know."

"Will a nuclear bomb fit into a shuttle? Could the bombs be exploded by a radio signal from 75,000 miles away?"

***

Sam had everyone meet him in the Captain's ready room. The Chief Engineer had finished his work.

"Gentlemen, except for 90 shuttles, all of the 1,800 hundred shuttles attached to *New Hope's* hull had the virus implant," he reported.

"The number of shuttle crafts in pursuit numbered in excess of 2,000," stated the Colonel. "The ships were not close enough to get an exact number. But I can now say with more certainty that the pursuit craft would get inside the 40,000 miles necessary to be able to use the virus to stop them. I believe the percentage of shuttles with the virus versus those without the virus would be about the same in each group. This means approximately 100 of Earth's shuttles would not explode.

"The shuttles could be destroyed all at once, or a select few at a time. We have no way of identifying each of the pursuit shuttles, so we will have to use a blanket signal. That means every shuttle attached to *New Hope* will have to have the virus disarmed quickly."

The Chief Engineer assembled a group of techs and started immediately. The question of how to stop the remaining one hundred shuttles was discussed.

"The Colonel told me we have 800 nuclear bombs aboard that could be remotely detonated," said Sam. "One bomb would easily fit in a shuttle. But there seemed not to be any other way to target the Earth shuttles.

The Colonel called up one of his best programmers and had him report to the Captain's ready room. The question was posed to the enlisted man. He thought for a minute.

"Yes, the shuttles can be controlled and started in the direction of the pursuing shuttles. That was about all that we can hope for, unless...."

"Unless what?" interrupted Sam.

"All the shuttles have a proximity alarm. When it goes off, the shuttle moves away. So why don't we change the shuttle's reaction to the alarm and have it close in on the nearest pursuit shuttle? The bomb's fuse can be integrated into the shuttle's radar and programmed to explode when it gets close enough," the tech explained. "You would have to wait to detonate until all the shuttles with the virus are destroyed."

"There is one more thing we need to discuss," said the Colonel. "It now looks like the encounter is going to happen just short of Mars' orbit. Any Earth shuttles that passed the halfway point between the moon and Mars probably would not have enough fuel to return to Earth. Surely the shuttle captains realize this? The shuttles are passing the moon now."

They started preparing 225 *New Hope* shuttles for launch, changing the software, loading the bombs and hard-wiring them into each shuttle's computer system. Putting the bombs on board was going to require teams of technicians to enter the shuttles, but this could be accomplished without going outside. The bombs were going to take the longest amount of time to finish.

By now, everyone on *New Hope* was aware of the situation. All the preparations were proceeding on schedule and would be ready. The pursuit shuttles were approaching their point of no return and would soon be within 40,000 miles. *New Hope* would soon be passing Mars. The shuttles with the bombs onboard would soon be ready to launch. The virus in the remaining shuttles had been removed.

The command to activate the virus was given and nothing happened. Sam and the rest of the command staff could not believe what they were seeing. Then, Earth's shuttles broke radio silence:

"Stop your engines and prepare for boarding, or you will be fired upon. Your virus was disarmed. We can easily approach and destroy your ship. You are defenseless. You have one hour to accept or be destroyed."

Everyone was completely caught off guard. Sam was the first one to recover.

"I don't believe them," he said hoarsely.

None of the shuttles had ever been armed! Just then, a small particle beam was observed going by, just missing *New Hope*. Strangely, it looked exactly like *New Hope's* propulsion particle beam, only a smaller version.

They had to put together a new plan of action very quickly.

Why were we given one hour to agree? Sam thought.

The Colonel must have been thinking the same thing.

"I think I know what is going on. We need one hour to recharge a rail gun. The shuttle was not large enough to have a breeder reactor on board. Therefore, it had to charge up a capacitor to generate enough power for the shuttle's electrical system to fire the particle beam," he said.

The lowly tech was the first to figure out a workable plan. It was true that *New Hope* was not armed, but one hundred shuttles had been turned into guided missiles. The shuttles were immediately launched and positioned so that the distance between them and the blasts would not overlap.

"The bombs had been designed for use in a planet's atmosphere for the blast and shock wave effect," announced the Colonel. "This did not work well in the absence of an atmosphere."

The bombs were reconfigured as neutron bombs. This would produce a debilitating electromagnetic pulse when detonated. This would provide an enormously larger area of destruction. The shuttles had been designed to operate between the Earth and the moon, a distance of about 238,000 miles. Therefore, they did not have radar that was designed for extremely long distances, and it was hoped that the pursuers would not pick up the awaiting bomb-loaded shuttles. The reprogrammed proximity avoidance system would make the decision when to detonate.

Sam and the Captain were having a deep conversation about what to do next. Their decision was to wait and see whether the ambush worked or not. They would know in about three hours if the ambush was successful. His next move was to stop the pursuers from firing any more particle beams at them. The group dispersed.

Sam had several nagging questions. He called in the orderly about an hour later and asked him to request the Captain and the Colonel come back to the ready room.

"How far did the particle beam miss, Colonel?"

"Close enough that our onboard collision sensors detected it. The sensors only have a range of about two hundred miles."

"Does anyone think they meant to get that close?" asked Sam. "I do not think they have the ability to fire accurately. Before we launched, the technology had not been developed yet to aim a particle beam even 7,500 miles. The target and firing point are both moving."

The Captain suggested slowing down so their pursuers would not fire again. Everyone but him agreed this would be a bad idea.

The three men decided to continue and depend on the bomb-loaded shuttles and to have everyone come to the ready room in two hours.

"The targets are within range, and now is the time to eliminate them," the Colonel reported.

Everyone agreed, and the order was given to activate the shuttle bombs. The ships continued to close the distance to the waiting shuttles. About five minutes had passed when a radio message was received.

"The next shot will not be a warning shot. The particle rail gun is ready, and we will destroy you..."

The commander never got to utter another word. They had reached the awaiting shuttles' kill zone. Assessing whether any ships have survived will require some time. Over the next few hours, it became apparent that the main force had been destroyed or disabled, and they were no longer a threat.

New Hope was now crossing Mars' orbit and continued to accelerate at 1 g. About a dozen ships at the back of the formation didn't seem to have been affected, and they continued. They appeared to be making a course change. They were heading for Mars.

It suddenly was clear to everyone how the surviving ships were going to return to Earth. They were going to use Mars' gravity to slingshot around the planet, and if successful, they should be able to make it back to Earth.

New Hope received one last message. "We will find you someday, and you will have to pay for what you have done; Earth will survive, but you will not. We have…"

The captain called the communications officer. "Is this how the last message ended?" he asked.

"Yes, Captain, it ended with 'we have,' but it ended in the middle of a sentence," he reported. "A computer analysis of the cadence of the words in the last sentence indicated the speaker was cut off. There was one other strange aspect about the message, Captain. It contained a string of highly encrypted code that just seemed to disappear into *New Hope's* computer system. We were not even able to record it and were not able to locate it for analysis."

"Let me know when you solve the riddle," the Captain said.

CHAPTER 5

The Early Years

Samuel Otto Bernard, Chief of Operations, sat in the Captain's ready room all alone with his thoughts. He looked back over the last 30 years. So much lying, deception, and murder, and he felt personally responsible for most of it. The Project had pitted all the distinct groups on Earth against one another on purpose. The plan had worked perfectly. The groups had spent the last 40 years trying to destroy one another and this diversion had allowed Project Migration to succeed.

Sam believed Earth was headed back to the Stone Age. Long ago, he realized everyone on this ship had only one goal, and that was their own survival.

***

New Hope had been built for an estimated 219,000-day journey, which was roughly 600 years. This equaled about 20 generations, using 30 years per generation.

The ship was a sphere, accelerating at 1G. This allowed all living things to continue to live and grow the same as they had on Earth. All actions and processes in the mechanical and chemical world would be the same as on Earth. This eliminated the necessity to adapt everything to zero gravity. The ship could survive brief periods of zero gravity – after all, the ship had been built in space. Also, the cloning process had a high rate of failure in zero gravity. No one knew why.

The ship's engine did not have any moving parts. The initial load of uranium fuel for the ship had a half-life of 5,000 years. The necessary DNA to clone all living creatures from Earth was in storage. The same held true for plants, especially edible plants.

The front section of the ship contained the bridge and the operational crew. Moving toward the rear, the Colonel and his marines were located just behind the front section. All the other crew members occupied the third section. The rear of the ship contained the ship's engines and uranium fuel for the shuttles.

The ship had been divided into four sections for security reasons. Movement from one section to another required going through various checkpoints. This would allow for a trouble spot to be isolated and dealt with. The ship had been constructed with full knowledge that anything could happen in 600 years.

There was a maze of secret hallways and stairs that all originated from Colonel Hooker's security section. There was not a section of the ship his men could not access in secret. Only a limited number of people knew about the hidden passageways, as the information was shared solely on a need-to-know basis. The nonessential workers involved in this part of the ship's construction had long since disappeared.

Food production was going to be closely monitored. The large amount of fertilizer needed had been solved somewhat by crop rotation and the use of the ship's treated sewage. The Project had been doing research back on Earth, and the introduction of genetically engineered crops would prove to be a tremendous help.

The vessel was expected to bring several thousand acres into cultivation, in addition to establishing hardwood forests and some pastureland for livestock. The availability of 24-hour artificial light for photosynthesis hopefully was going to shorten the time needed to grow any particular crop by approximately half.

Sam realized better than anyone else that the smaller the number of crew, the easier running the ship would be. However, reducing the crew size to the target number of 2,100 by placing several thousand people in cloning storage was a challenging task. Sam was going to have to consult Dr. Harris to get her input.

Dr. Harris and Sam had limited familiarity with each other, having only spent a few hours together over the past two decades.

Come to think about it, he thought, *she had always avoided any small talk and opted out of every meeting possible. I have to research her background before seeking her counsel.*

Sam pulled up everything he could find on Dr. Harris. It proved to be not much. Sam knew he was going to need expert help if he was going to do a deep search about Dr. Harris. The young computer tech that had helped destroy Earth's pursuit shuttles came to mind, but he pretended to not remember his name, so he contacted the colonel and asked for the programmer to report to him in his ready room. The tech showed up in about 30 minutes.

"Good day, sir. I am Thomas Hacker, Warrant Officer, First Class. But everybody calls me Tom," he said, introducing himself to Sam.

He was responsible for maintaining all shipboard computers, as well as overseeing computer security. Tom was 38 years old and was one of those people who seemed suspiciously older.

"Good day, Tom. I want you to tell me about the computer security on the ship," asked Sam.

"Well, sir, the computers on *New Hope* contained about the same amount of stored data as all of Earth. I'm not sure if they were indeed equal because no one knew how much Earth data had been available to load," he explained. "Nevertheless, all of the data they could find on Earth had been loaded aboard *New Hope*. The data had been compressed and any redundancy discarded.

"Access to the ship's computers is compartmentalized, and each section has three levels of access. All computer users can only access their section's data and at their appropriate level."

As their conversation continued, Sam was finding out more about the clandestine design of their ship. Sam soon realized that he should have spent more time and played a larger role in the ship's design. He was going to need someone close who understood all facets of the ship.

I'm going to offer Tom a promotion, Sam thought. *But first, let's see how he handles his first assignment.*

"Tom, I want you to search and gather every shred of information you can find about Dr. Harris and report back to me," he instructed.

Sam's world had slowed down dramatically. There was day after day when he was not involved in any of the day-to-day operations of the ship. The ship would soon leave Earth's solar system and enter open space.

***

Dr. Harris knocked on the Colonel's door.

"Enter," he called.

"Oh, good day, Dr. Harris… Emily," he said as Dr. Harris entered. "How can I help?"

"Colonel, only a small number of crew members volunteered for storage to be cloned at a later time. I don't know what to do."

"Mmm... It's very early in the voyage. I don't want to use physical force at this stage," he mused. "Let's hear what Sam, the Chief of Operations, thinks."

They met in Sam's ready room alone. None of the three thought it was a particularly clever idea to take down crew members out in the open. There were an estimated 1,800 people who needed to be put in storage. The ship could support a crew of 2,100 indefinitely and several thousand more for short periods. After some discussion, they had a plan.

"So, we will first see how big our problem actually is," concluded Sam. "We will actually count the number of crew members awake and see how big of a problem putting them in storage is going to be."

However, this proved to be easier said than done. None of the ship's six departments had an accurate count. When the ship left early, workers who should have been returned were still on board. So, Sam instructed each section chief to count everyone under his command and get a count of workers who were not supposed to be aboard after launch.

The total number was a shocker; 4,400 people were awake. That meant 2,300 people would have to be dealt with, consisting of 400 crew and 1,900 workers. The workers were expendable and were told to volunteer for storage and report to Dr. Harris, or they would be forced into storage and never be cloned. Worker volunteers would be put on the rotation list. The option to wake the workers if needed was a plus for the ship.

That solution could be accomplished quickly. The 400 extra crew members could be absorbed easily. This would allow Sam sufficient time to use the normal rotation over the next 30 years to store more individuals than those who were awakened.

***

Sam had been waiting patiently for Thomas Hacker to finish his investigation of Dr. Emily Harris. Months had passed, and he had not heard a word from Tom. It was time to find out what was going on.

He called the Colonel. "Good day, John. Won't you please ask young Tom Hacker to report to me as soon as possible?"

A few minutes later, Tom appeared.

"Good day, sir. How can I be of assistance?" he asked politely.

"What have you found out about Dr. Harris? It's been a few months since I asked to find information about her," Sam asked.

"Well, I didn't think it was that urgent," Tom said. "And I want to do a proper job."

Sam sensed a difference in Tom. Something had changed.

"Well, what did you find?" prompted Sam.

"I don't quite know. It is puzzling. There was nothing in the ship's computer that mentioned Dr. Harris's existence prior to her coming on board the Project as the department head of Medicine and Science. The information available from her application was on file, but the information from the background check was missing," Tom explained. "She did not exist; I couldn't find any records."

"Now, what does she want to hide?" asked Sam.

"Either Dr. Harris had stopped any mention of her from being uploaded or had placed a virus in the ship's computers to prevent anyone from being able to find anything about her before she joined the crew," said Tom. "The data must have been uploaded and placed in an ultra-secret file. I think this is the most likely scenario."

Sam decided that before giving Tom instructions on what to do regarding Dr. Harris, this would be a good opportunity to talk about his future position as assistant to the Chief of Operations.

"Tom, you did an excellent job. I want to offer you a promotion to Assistant to the Chief of Operations," said Sam as he took out a bottle of scotch from the cupboard. "Here, have a seat and a drink."

"I still have some work to do on my shift, so I have to decline the drink," Tom said as he sat down.

Sam went over Tom's new duties. The first and most important duty was to report any and everything, no matter how big or small. He would decide what steps to take.

"Remember, even the smallest details were sometimes the most important," Sam stressed. "You will be promoted to the rank of commander, putting you equal to all department heads. Only the ship's Captain, the Colonel, and I will outrank you. You will move into the vacant quarters next to mine. I will instruct Colonel Hooker to provide all the necessary codes so you can enter all parts of the ship. And you can decide when you want to be cloned."

Tom was overwhelmed by this turn of events. He smiled broadly.

"Well, what do you say?" asked Sam.

"I'm very grateful for the opportunity, but…"

"What?"

"If anyone tried to stop me from doing my job, what recourse would I have?" asked Tom.

Sam could see where this was going. "Tom, you can use deadly force at your sole discretion in the performance of your duties. I will see to it that you are armed with a concealed weapon."

All the ship's weapons were locked up in the armory. Colonel Hooker was the only person openly armed with a side arm. He thought he had the only weapon outside of his armory. However, Sam had made sure he could arm a full company of men, if necessary, by hiding weapons in a secret closet in his quarters. So, providing Tom with a weapon would not be a problem.

"Then my answer is yes. I would love to be your assistant. But first, I need time to find someone to replace me in my current job," replied Tom.

Sam had done all the research himself. Tom did not have a single living relative; he hardly knew anyone and had no close friends on the ship.

Sam went to the Bridge to inform the Captain about his new assistant.

"Well, what can I say, Sam," said the Captain, slightly taken aback.

"Naturally, you would have access to all the information Tom might uncover, Captain," explained Sam.

He failed to mention the other aspects of Tom's job, mainly Dr. Harris.

***

As time passed, things kind of mellowed out. The ship had settled down into a lazy routine. Sam still had not made any progress in resolving the issue around Dr. Harris. Tom had become quite proficient in rooting out trouble before it happened.

The operation of the ship was going better than expected. Anastasia, as Head of Food Production, was able to get food production up and running. The genetically engineered crops had worked perfectly and were growing better than expected. The one thing that was feared the most did not happen; the flora and fauna that were aboard were largely disease-free.

This was true for the crew also. Only a half-dozen deadly viruses and bacteria had managed to make it on board. Anything contaminated had been put in isolation and then released into space. Sick crew members were put in cloning storage, and their bodies were destroyed. The isolation of the crew into the four different sections had not proved to be a problem yet.

Sam and Tom tried to meet at least every 180 days. At one of the meetings, Tom announced he had collected all available information on Dr. Harris. Sam was elated.

"All records indicate Dr. Harris is 107 years old. Her current DNA indicates her present body's age. Dr. Harris was involved in the research on transferring a person's consciousness into a new cloned body almost from the beginning. So, she would have had ample time and opportunities to clone herself.

"The ship's records indicated she was 52 years old. She had cloned herself sometime in her mid-fifties, 10 or 15 years before you became Chief of Operations. There is a large file I am unable to completely access concerning the ship. The info about Dr. Harris came from uploaded data from Earth and data created aboard the ship since its launch. I believe Dr. Harris might have a large secret group of the crew involved in her plans – they called themselves the 'Fleas.' I will keep on trying to access the hidden data," Tom said.

Sam stroked his chin, thinking about everything Tom had told him.

"Can I enlist help, Sam?" Tom asked. "There are some techs who are better hackers than me."

"Maybe," Sam replied. "But I want to meet and interview any potential candidates before bringing them into the group and sharing information."

"You see, since we were dozens of years into the trip, we will have to accelerate the reduction of the crew to the minimum number," Sam explained. "Identify all the Fleas first. We'll put them in storage first. That will hopefully reduce the threat."

As soon as Tom left, Sam sat back down and began to go over these latest developments. He first went over the cloning process. The cloning process for a live clone was as follows: the new body could be grown to whatever age was desired by harvesting the appropriate DNA. This could be accomplished in a maximum of six weeks. The lab techs explained the problem during cloning; removing physical racial characteristics was slow, but the procedure would only get faster.

Once the new body was ready the transfer could be accomplished by almost anyone. The transfer involved placing the old body next to the new one and placing harnesses on both heads. Then it was only a matter of flipping switches and waiting for the data transfer to finish, usually about 30 minutes. Once the new body's brain took over, you just disconnected the new cloned body from the heart-lung machine, and you were finished. Most newly cloned people did not require any further assistance. There was one strange anomaly; almost nobody ever remembered their previous body.

When a person was to be placed in cloning storage, the procedure was extraordinarily complex. Few people on board had the training and experience to operate the equipment to preserve the data stored in a person's brain. The old body could be rendered to its basic elements and used at a later date. The data content of a human brain could only be retrieved and used if it were captured before death. This was most important because a death from an accident, sickness or murder would end that person's existence.

Sam believed a person if left alone, could clone themselves if a new body were available. It was during this time that racial characteristics were being cloned away, and the crew was beginning to all look alike. The shrinks hoped this would eliminate CRT (critical race theory) friction between different races if everyone looked alike. The same fate awaited the alphabet people's different gender identities in the hope that sameness would solve the problem they sometimes presented.

Sam decided to visit with Tom every inch of the ship. Tom was the only person on board except the Colonel, who had complete access to all the computers on board. Sam earlier had not believed he would ever need the passwords, but now he did. The discovery of Dr. Harris's deception had shaken Sam more than he had realized. He was going to have to be careful about which people he chose to be in his inner circle from now on.

CHAPTER 6

The Balkanization of the crew

Tom already visited every square foot on the ship, and now, he and Sam would start at the rear of the ship. They decided to use the secret passageways that originated from the security section. This would make him only the second person who had actually traveled throughout the ship using the security passageways.

"Why don't we use the normal routes?" Sam asked.

"It's becoming harder and harder to pass through the checkpoints. The delays are becoming longer and longer by the day. The checkpoints were designed to only be used in case of security needs, but the different sections were beginning to take control of the checkpoints with their own people. The only security people who were authorized were Colonel Hooker's men," Tom explained.

Sam decided to go to the rear of the ship to the engine room. There was a corridor that went straight from the bridge and security to the engine room. It contained elevator cars to allow multiple users at once. The trip, at normal speed, lasted about 15 minutes.

Wolfgang Schmitz, the Chief Engineer, was waiting for them. The tour of the ship started, and four hours later, he was still not through. They entered the Maintenance section.

"Sam, this is Qiao Hui, the head of the department," Schmitz said.

The public corridor leading to the next section was blocked off for safety. Sam was beginning to realize something was horribly wrong.

If the rest of the ship were blocked off this early in the trip, we would never survive the journey, Sam thought.

Sam decided to meet with the Colonel first and see what his take on the developing situation was. Sam needed at least one ally to help him assess the present situation.

The colonel began to describe and explain in great detail what was taking place at this very moment.

"Because of how the different sections were isolated from each other, restricting ingress and egress to the various sections was becoming increasingly popular. Hoarding food and supplies was accelerating, which could mean only two things. Each section was either stealing from each other, or they were trading among themselves using whatever valuable commodity each section had under its control," explained the Colonel. "I have been using the ultra-secret surveillance cameras located in the most critical areas of the ship since day one. This included everyone's private quarters."

Sam realized the Colonel had recorded every word spoken on the ship since its launch and had backed up video to boot. Choosing the Colonel if his intentions were the same as his own was good news. Everyone expected surveillance cameras, and they had been placed so people could get out of the cameras' view. Secret cameras were placed to take advantage of this behavior.

"Colonel Hooker, how many of your men know about the secret camera network? And how many are authorized to view the footage?" Sam asked.

"I'm the only one who has the passcodes to access the system," Colonel Hooker replied. "As far as I know, nobody else even knows about the secret cameras."

There were tens of thousands of cameras on the ship, and they were all integrated together, but the security people could only access the cameras in plain sight – never realizing that the other cameras existed.

"Colonel, please prepare a thorough timeline of Dr. Harris's activities from the beginning," Sam asked.

"I have already created one. The report is ready to view now if you want to."

Suddenly, Sam noticed the Colonel watching him intently, waiting for his reaction to this new disclosure. His hand was on the side of his arm.

Sam realized he had been outmaneuvered. Now, he had no choice but to enlist the Colonel's help. He now knew full well he would never know the full extent of the Colonel's knowledge. This opened up all kinds of problems, trust being number one.

Sam and the Colonel sat down to look at the prepared report. Dr. Harris was among the first research scientists to sign on. She was given the credit for developing the last piece of the cloning puzzle, which was the transfer of a person's conscience mind into a new body. This first took place about 50 years ago. The corporation had purchased the knowledge and personnel needed to perform cloning. They were all brought aboard and began to construct the necessary facilities to clone the crew as necessary during the trip about 30 years ago.

Records from Earth about Dr. Harris's age and birthplace were sketchy and sometimes conflicting. All the information pointed to Dr. Harris as being 107 years old, and this agreed with Tom's assessment. This could mean only one thing – she had cloned herself in secret.

Sam and the Colonel began laying plans to stop Dr Harris. The Colonel was aware of Dr Harris's secret organization.

"Tom had also discovered them, and we've named them the Fleas," disclosed Sam. "Tom and I can only guess Dr. Harris's intentions."

They continued to watch the Colonel's report. He really did have a complete picture. The Fleas extended throughout the ship and were in every section. Dr. Harris communicated through a secret Wi-Fi system using an encrypted signal. Their numbers were relatively small, only a few hundred of them. The Fleas were in almost every leadership position, but they still had a way to go. Some of the ship's departments remained unaffected. Dr. Harris apparently had a plan to slowly take over the ship bit by bit. The bridge, crew and security had not been infiltrated yet, but there had been one Flea in the bridge computer department.

"What had happened to that Flea?" Sam asked.

The Colonel looked at Sam. "You transferred him and made him an assistant to the Chief of Operations, holding the rank of Commander."

It suddenly hit Sam; Tom had been cloned recently. That was why he seemed older than his age, and that was because he was older. That also explained his turning down the drink. For many people, it took several years after being cloned to acquire a taste for alcohol again. This was devastating news.

They continued with the report. Dr. Harris had realized years ago that when you cloned someone, it was an opportunity to control that person's behavior in the future. The same idea educators had with children: plant ideas and behavior traits early, and they will turn out to be the adults you desire. The Fleas under Dr. Harris's direction would have their people in control of the key positions on the ship in a few years.

Sam and the Colonel did not think the Fleas knew about the secret camera network. Until they did, Sam and the Colonel would know about their every move.

"Well, let's not share any of our intel with Tom," said Sam.

"Agreed," replied the Colonel. "I'm stumped about his recent cloning and the lack of information about his background."

They decided to spend the next few weeks gathering information and creating a plan, then meet again in a few weeks.

Why did Tom tell me about the existence of the Fleas in the first place, Sam wondered.

Sam lost track of time going over a plan of action. He decided to have Tom and the Colonel both try to find out if there was any way to clone people and put people in cloning storage secretly.

There had to be redundancy built into the cloning process, meaning more than one laboratory set up to clone, but where it was located and who controlled it, Sam did not know. The Balkanization of the ship's crew was happening before his eyes. Soon, each of the ship's sections would seek control of the ship.

Sam's problems were beginning to be plural. The first order of business was to get the reports from Tom and the Colonel. He decided to ask each of them to report in writing, the old-fashioned way. This would allow Sam to study each report privately and compare them.

***

Months went by, and finally, the reports arrived.

Sam read Tom's report first. Tom believed that a second cloning laboratory was being run by Colonel Hooker's security section. They have managed to isolate and place in cloning storage all the Fleas in the security and bridge section. Tom was sure Dr. Harris was also aware of this. He also explained that the Colonel must have in his department people who were capable of running Dr. Harris's medicine and science section.

Tom now believed all the computers on the ship could be monitored by the security section, but not all could be altered without leaving a trail. Monitoring the ship's computers could be accomplished without leaving a trail. The report had taken Tom months of tedious work online, and several times when he happened to go back, he noticed changes in confidential and ultra-secret files. He had an eerie feeling that someone was observing his searches.

Sam pulled out the colonel's report and opened it up, but the pages were blank except for the words "IT'S TIME FOR US TO TALK."

Sam went to the Colonel's office. He almost fainted when he saw the Captain and Chief Engineer also in attendance.

"Good day, Sam," greeted the Colonel. "Please join our conversation."

Sam sat down. The Colonel was in his element and immediately took charge of the discussion. Sam immediately realized he was the outsider here; these three had obviously been having these strategy sessions on a regular basis. They had avoided discovery by being extra cautious. Sam realized this was going to be an interview.

The Colonel produced a voluminous written report. The words "Samuel Otto Bernard S.O.B." were typed on the cover. Sam noticed the Captain and Chief Engineer also had a copy each.

"Sam, have you ever been cloned before?" asked the Colonel.

"No."

"Why did you pick Tom to be your assistant?"

Sam decided to play it straight.

"I needed some secret computer searchers, and then I remembered the young computer tech that had been extremely helpful and played a big part in destroying Earth's fleet that pursued *New Hope*. I asked the bridge for help getting Thomas Hacker reassigned."

"Did you know Tom had been cloned right before the ship launched?"

Sam thought for a minute about his answer.

"No, but you did, Colonel Hooker. Otherwise, you would have picked an older, more experienced tech. How old is Tom now?"

"His body is five years old, and his mind is 75 years old," replied the Colonel. "I secretly cloned Tom a second time to remove his genetic propensity to be a Flea. But he remains a Flea and is now a spy for this group. We had always through that Dr Harris was the originator of the Fleas. If she is not, then who is?"

Unanimously, they all answered, "The chair of the board, Brock."

"What are you going to do about Brock?" asked Sam.

"We have already taken care of Mr. Brock," said the Captain.

This news startled Sam, but he had to know. "How?"

"He was falsely diagnosed with a brain tumor. His symptoms were induced by drugs in his food. Brock realized he had to be cloned before he died. We used the security cloning laboratory but put him in storage instead of completing the cloning process. The story was Brock died before the cloning procedure could be begun," replied the Captain.

The justification for the security cloning laboratory had originally been to clone wounded security officers as fast as possible. This eliminated lengthy recovery time. But now, Sam and the others realized that the cloning lab was going to prove particularly useful in solving the Flea problem.

***

Tom was asked to attend the next meeting which was scheduled for 190 days. Sam had noticed people slowly beginning to stop measuring time in months, weeks and days of the week and began to use the number of the days remaining in the voyage. They all met in the Colonel's tactical planning room.

Tom had used the last 190 days to examine Dr. Harris's moves, and he had prepared several possible courses of action to take.

"The Fleas have to be stopped, and Dr. Harris has to be dealt with. We need to establish the extent of the Flea infestation," said Tom. "Their long-term plans were unknown to the average member – at least, that is what I gathered."

"As far as we know, the bridge section, security section, engine room and maintenance section were free of Fleas," the Colonel said, glancing at Tom. "Almost all our information came from Tom, a former active Flea. The two sections containing the most Fleas were Science and Medicine and Food Production."

These areas occupied the most space. They were located in the middle section of the ship. Thousands of cubic yards of Earth's finest soil were on board and placed on several decks located at the ship's equator, or approximately 314 square miles on each deck near the ship's equator. They were occupied by all of Earth's flora and fauna. The two sections in question were almost twice the area of the remaining ship, with Anastasia's food and resources by far the biggest section. There were hundreds of square miles of forest that security had not bothered to explore.

"But let's infiltrate the two sections," suggested Sam, being the most cautious. "Then we'll find out what we are dealing with if we decide to make a move out in the open against Dr. Harris and Anastasia."

"Have you read the fine print buried in the middle of my report?" Tom asked.

Everybody in the group shook their heads to indicate "No." They weren't used to reading printed pages any longer.

"Turn to page number 71," Tom instructed.

After reading, the group members looked up.

"So what?" the Colonel said.

"Dr. Harris had brought aboard copious quantities of all the medicines needed to cure Earth's diseases," Tom explained.

This time, Sam spoke. "So what."

"Not a single person on the ship in the last five years had been sick enough to require extensive medical treatment. Look, I believe, as do all the medical people I had made friends with who were also Fleas, that the extraordinary precautions taken to eradicate all the pathogens that caused disease and death from being brought aboard before launch had been successful. So, there is no need for medicine in such copious quantities," Tom explained.

There was a shocked silence. Everyone was stumped by this revelation. A decision to have the Colonel explore the areas in question was unanimous. No other action would be taken until they have some hard intelligence.

Sam went back to his quarters to enjoy peace and quiet. He had more information to process. A lot had changed in the last year or so. Each section of the ship was becoming more isolated every day. It was almost impossible for people to move about the ship freely between sections. Groups were bartering between themselves for their everyday needs. This was beginning to affect the day-to-day operation of the ship. Soon, the Balkanization of the crew was going to endanger everyone aboard. All the same, problems that had plagued humanity were taking hold again and had to be stopped.

I am still feared enough, Sam thought. *If I produce a plan, I can regain control.*

CHAPTER 7

Slow Times – the Middle Voyage

Sam had spent years, sometimes with Tom, exploring the ship, spying on everyone, trying to produce a plan. One area, at the center of the ship, he had not physically explored: the large area containing farms for food production and large wooded areas that were not being used yet.

All the surveillance cameras appeared to be working properly. The status report Tom produced indicated all cameras were working, and every square foot was available for observation. But somehow, Sam did not believe his own eyes or the computer information. He finally decided to contact Tom so they could meet.

"Tom, I don't believe everything the information we're getting," Sam told him.

"I know. I have the same feeling. The cameras are hiding or omitting large areas which were supposed to be devoid of any continuous human activity."

They agreed to plan a scouting trip, and until final plans were made, Tom would do a complete examination of the ship's company. Then, they would figure out their next move.

Sam knew he was going to have to produce a long-term solution so that he could convince the others that his plan was the right one. They were years away from their planned destination. One thing Sam knew for sure; upon arrival he must have complete and unchallenged control over the entire ship. He soon realized he had all the time in the world to make his plans. Sam believed that Colonel Hooker and his marines were the key to maintaining control.

He tried to contact the Colonel only to discover that he was recovering from a recent cloning. Then it hit Sam.

I am almost 90 years old! he thought. *If I die now, I will cease to exist.*

CHAPTER 8

Sam's Cloning

The ship had two facilities that he could use, but which one would give him the best outcome? Would it be Dr. Harris or Colonel Hooker? Dr. Harris cloned hundreds of the ship's crew every year; Colonel Harris only cloned his security people and a few people from the bridge, maintenance, and engine room sections.

Sam decided to do some investigating himself. He wanted to spend some time meeting and talking with people who had been cloned since liftoff. He would save the crew members Colonel Hooker had cloned till last. Sam realized this was going to take months, maybe years, to accomplish without raising suspicion.

The more people Sam met from the various sections, the more he suspected something was changing, but he could not put his finger on whatever was changing people's behavior. Then, it dawned on him: the crew was slowly changing into two diverse groups.

Sam decided to make the necessary arrangements for his cloning using Colonel Hooker's facilities. Secretly, Sam gave himself every psychological test available and put into storage all his physical specimens for comparison after cloning. Finally, it was the big day.

Everyone at the lab was all smiles as the complex preparations were performed. Sam was scared he could not get it out of his mind that what his eyes were seeing might be the last thing he would ever see.

When Sam woke up, he felt exactly as he had before. Nothing had changed, as far as he could tell. The protocol called for him to get a thorough physical exam. When they finished, he returned to his office and quarters.

Upon walking into his quarters, he went straight over to his computer. The average time for cloning was now 12 to 15 hours; any more time and there would be a reason. When he left for the lab, the clock on his computer indicated 05:10 day number 29,900. Now, the clock showed the time as 10:27 day 29,904. He had been unconscious for 96 hours longer than the average person.

Why did he remain unconscious for so long? Nothing had been discussed with him about any plausible reason for the extra time. This could only mean one thing; the people who were present when he was awakened were the ones responsible. Clearly, there was a reason, but what was it? They could have just allowed an accident to happen, and that would have ended him.

When Sam designed his quarters, he had taken precautions for his safety. The quarters in an emergency situation could be sealed off from the rest of the ship, similar to an old-fashioned safe room in a house. He decided now would be the right time to find out exactly what had happened during his recent cloning.

He sat down at the computer and constructed spreadsheets comparing the old Sam and the new Sam. There were areas, both physical and mental, that just did not look like they were indicative of the same person. Something major had happened to him, and he was not quite sure how the changes were going to affect him mentally and physically.

At this point, he did not know who to trust. He decided to get some rest. Thereafter, he would contact Tom.

He was awakened by his quarter's perimeter alarm; someone was trying to gain entry. Sam looked at the camera over his door. Tom Hacker was using the code Sam had supplied, but it would not work when his quarters were in lock-down mode.

Sam realized he only had a few more seconds left to turn off the alarm, or Tom would be killed after three unsuccessful attempts to open the door. A short blast of poison gas was about to be unleashed and would kill anyone within about ten feet or so of the door. The gas was mixed at the nozzle and would be inert and harmless after about 20 seconds after release. He had to hurry.

Sam opened the door and let Tom into his quarters, not telling him about his narrow escape from the gas. They had some catching up to do.

"You go first, Tom," said Sam.

"Unfortunately, I couldn't get into the middle levels of the ship. Those areas are controlled by Dr. Emily Harris. So far, she had been able to control all access, and only people she approved of were allowed to enter. I believe every cloning by the medicine and science department is implanted with the right genetic makeup to become a member of the Fleas. Her organization was growing daily," Tom said.

Sam was just about to start his report when Tom continued.

"I've detected another very disturbing development. The number of people who want to remain in cloning storage until the end of the voyage is growing daily."

This was unexpected news.

"What does that mean?" asked Sam.

Tom was also at a loss to explain it. They discussed what might be happening that they were missing. Tom had spent almost all his time out among the crew in all the different areas of the ship.

"How much time have you actually spent among the crew since launch, Sam?" asked Tom.

"Not much, I have to admit," replied Sam. "At the very beginning, I went to all areas of the ship. The crew was always mostly upbeat and optimistic about the great human migration. But the voyage is having a strange effect on everyone on board... and not in a beneficial way."

One of the corporation's most obvious mistakes was that the human knowledge that was loaded into the ship's computers lacked any mention of religion. This had been done on purpose. The social engineers had hoped to prevent friction that would eventually lead to violence among the crew. Instead, the effect sped up the process. Religion could have been the glue that held the crew together.

The six distinct ship departments no longer permitted free access throughout the ship. Colonel Hooker and his men were still able to move about the ship, but it was only because they had all the guns. Dr. Harris and her cloning techs also had free movement, but only when escorted by the security people of the department being visited. Everyone else was trapped in their respective departments.

The one person who was making progress was Dr. Harris; as the Fleas grew in numbers, she became more powerful.

At this point, Sam decided to change the subject. "Tom, have you noticed anything unusual about my recent cloning?" he asked.

Sam was looking closely for Tom's reaction to the question. Sam realized that the person known as Tom was well over 140 years old, but his body was only sixty or so. He realized he was not going to outwit Tom. Sam was also well over 140. Suddenly, Sam realized Tom was on his side.

Sam had given Tom a heads-up about his upcoming cloning. He was not able to be there at the start, but Tom had long ago hacked every computer system on the ship so going over Sam's cloning data had been easy. He soon realized that Dr. Harris could do the same, and she had interfered in the process. Tom enlisted the help of some of the security department's technicians. They proceeded to do a do-over and needed an extra sixty or so hours between cloning. Tom removed all the changes Dr. Harris had made. All his changes were hidden. The first cloning was on record, and Dr. Harris would think she had been successful.

"Sam, you have to tell Colonel Hooker," Tom insisted. "It was he who asked me to check on you. Are you up to accompany me to Security to see him?"

Colonel Hooker was pleased to see Sam. But after hearing what Sam had to say, he became extremely agitated.

"I didn't know Harris had infiltrated my department to this extent," Colonel Hooker exclaimed. "But I'm very grateful for the information, especially since it might be about time for me to undergo another cloning!"

Sam, Tom, and Colonel Hooker all agreed it was time to discuss the ship's future in detail. First up were the behavioral changes showing up in people after their third or fourth cloning. Nobody seemed to be able to put a finger on it.

The Colonel had the most experience dealing with crew members after the procedure. He had often talked openly with his technicians, and everyone had multiple ideas as to what was happening but no proof. The consensus seemed to be the simple fact that everyone was going to be trapped aboard the ship for several hundred more years was having a devastating effect on the crew's psyche.

Sam, Tom, and Colonel Hooker all realized that no one actually had any idea, much less proof that would explain everyone's behavior.

CHAPTER 9

Mystery of the Middle of Their World

Sam tabled the behavior problem, but he wanted to revisit the past discussions about the large areas in the middle of the ship, not under the ship's security cameras. The areas were under the control of Dr. Harris.

The three of them all agreed she was the greatest danger to the success of the migration.

"First, I think we should send in a security team. They would travel and gain entrance to the middle area by using the secret security passageways created for just this very reason," suggested the Colonel.

The next question was how large of a patrol they should send. There were only about 100 security and marine personnel awake at the present. They did not know how many people they would find, if any.

"On the ship's original manifest, Dr. Harris had provisions for 800 crew members in her department," said Sam. "She may have shrunk the number, as had the rest of the department heads."

They decided that maybe they should send an overwhelming force of 100 men and women marines.

"I am going to need some time to wake up and prepare one hundred new marines," said the Colonel grimly.

The next and most delicate fact that needed to be discussed was armament for the patrol. The Colonel and a few of his men possessed only a few handguns.

"Colonel Hooker, should you make public there was, in fact, an armory on board, and only you had access?" asked Sam.

"Mmmm. I think I'll hold back that fact until right before the mission," replied the Colonel. "I'll have a timetable available as soon as possible."

"I think I have some interesting news. The computer algorithms I have been running secretly, since launch, shows that all the crew members involved in the disappearance of chairman Brock had themselves disappeared. This was about sixty crew members," said Tom.

"Is this just a coincidence?" asked Sam.

"I don't think so," replied Tom.

"But what does it mean, Tom?" asked the Colonel.

"I don't know."

***

The four of them (ship's Captain Dmitri Volkov, security Colonel John Hooker, computer expert Tom Hacker and overall Project Manager Samuel Otter Bernard) began what turned out to be a long discussion on how much information to share with the other department heads. They all were not sure how much Chief Engineer Wolfgang Schmitz, First Mate Oliver Whipple and Head of Maintenance Qiao Hui actually knew about the present state of the ship's crew. They all agreed that Dr. Harris knew everything that was going on in secret and was behind all the problems. They also, after much discussion, agreed Anastasia (Food Production) was also part of Dr. Harris's takeover scheme.

Each of the ship's departments had formed loose alliances among themselves. The act of people moving about the ship from section to section was becoming harder and was gradually disappearing. Sam surmised this had been the plan of Dr. Harris from the beginning. Her department was the only one that still could send her people throughout the ship. Everybody wanted to be re-cloned when the need arose. Every person cloned by her department added another Flea to her ranks.

This would explain the sizable percentage of people who were asking not to be re-cloned until the ship arrived at the new planet. Sam believed the people were never told what Dr. Harris had planned for them; the need for new Fleas trumped all else.

***

Hooker had estimated it was going to take about 400 days to wake up 100 new marines and another 200 or so days to train and equip them.

"However, to maintain confidentiality regarding the preparations, I need to find a way to conceal several complex aspects, such as increased food supplies and additional equipment for the plan to work," he explained. "Also, at this point, I haven't worked out how I'm going to hide all the preparations."

"Tom, use all the resources you have to find out all you can about the ship's interconnected groups," instructed Sam.

The gang of four hoped to be able to defeat the Fleas.

***

Tom began to assemble a group of people to begin the mission of collecting information from all areas of the ship to aid in the exploration of the ship's middle decks. Tom could see a long indoctrination was needed to bring his people up to speed. The group he had gathered came from all areas of the ship.

Finally, Tom asked them to accompany him to the bridge area, where he would explain the mission. The ones who agreed were absolutely blown away when they made their way to the bridge using Colonel Hooker's secret passageways.

The secret passageways seemed to instill a huge amount of confidence in them all. No one was absolutely certain how this diverse group would react to each other. The individuals arrived from various parts of the ship, each possessing distinct differences that could potentially hinder their ability to collaborate effectively. All the ship's departments were different, just as if they were all from different countries. This situation of mistrust and hostility developed since the launch occurred over 280 years ago.

Tom separated them the best he could by breaking them up into smaller groups. He convinced Colonel Hooker to provide his best men, and they became drill sergeants. This turned out to be a very workable idea and allowed the recruits to be separated.

The most limiting factor was the lack of space for training. The solution turned out to be, have three eight-hour training sessions, each 24 hours. They divided the men and women into three groups for training. Each group, when they started their training, each day had just gotten up and had breakfast. Essentially, each group operated on its own separate daily clock.

Sam and Tom had lengthy discussions about exactly what needed to be accomplished. They were plotting to control the ship, and they both believed others were doing the same thing. They agreed the most practical course of action would be to gather all the situational facts. The barter system that had developed between the ship's various departments was not fully understood by anyone, one person or department.

Hopefully, the new spies will be able to report back and provide the necessary information. This would give the gang of four a tremendous advantage. They would know exactly the individuals to eliminate when the time came. Tom would get back to Sam when he thought the spies were ready. The decision to keep the groups separated was, they hoped, the right one.

Colonel Hooker was experiencing problems of his own. He had decided to only wake men who had been put in storage before the voyage had begun. This he hoped would prevent any of them from having been corrupted previously by the Fleas. Because he had all the equipment to clone the body and store a person's consciousness, he hoped it would prevent the Fleas from infiltrating his security section.

CHAPTER 10

Thrust Reverse

The ship's Captain, Dmitri Volkov, had rarely used the ship's public address system, so everyone was quite surprised when he announced that all the available data indicated the ship was approaching the halfway point. This would require the ship to rotate 180 degrees.

"This is going to require the cooperation of everyone on board. You will have to set aside all differences for the ship to complete the maneuver and continue the voyage safely," the Captain ordered.

The exact details on how to accomplish this were not widely shared with the ship's company. Captain Volkov then explained it in the most user-friendly way he could.

In essence, the fact that the whole ship operated in a one-G Earth environment would make it difficult for the ship to avoid damage from even a brief period of weightlessness. But engineers believed that, in a pinch, it could be done. The ship's design engineers had produced a novel plan to accomplish the 'flip,' as it had come to be called lately.

The ship had 1,800 shuttles attached to the outer hull, and their thrust would be used to rotate the ship end over end while the ship's main engine thrust was pointing in the direction needed to continue. At this point, the shuttles would be shut down, and the shuttles on the other side of the ship would be used to slow the ship's rotation and stop on the course needed to arrive at the new home planet. The ship's main engine would then be brought back online and start producing one G of thrust, slowing the ship's speed down to zero over the last half of the voyage.

The last hurdle was going to be matching the new home planet's speed and direction and then achieving orbit. Two planets were orbiting around one of the suns, and one planet orbiting around a second sun. They would make a choice after they arrived and then officially name the new home planet.

"All department heads, contact me as soon as possible," the Captain ordered. "I'm going to talk to you as a group, but first, I need to talk to each of you separately. You should all bring one assistant," he ordered.

Then, Captain Volkov quietly contacted Sam, Tom and Hooker and asked them to come to the bridge in 19 hours.

***

They assembled on the bridge, looking at Volkov expectantly.

"I have an idea. Maybe we can use the flip as an excuse to deviate from the norm," he said.

"How soon is the flip going to happen?" asked Sam.

"And how long is it going to take?" added Hooker.

"Let me introduce the Chief Navigator," said the Captain.

It turned out to be an AI. The AI began the briefing. The maneuver could start anytime. Then, the AI explained as only a computer can.

"Do not start the rotation until the entire ship has returned to zero-gravity conditions, meaning the same as when the ship was under construction. Everything has to be secured in place. Water or any liquids has to be contained in sealed conditions."

Chief Engineer Wolfgang Schmitz had never been part of the gang of four. His department was going to be responsible for all the work required to prepare the shuttles. First Mate Oliver Whipple could be a problem; he was responsible for all the ship's preparations.

"Captain Volkov, I think you should approach the two of them separately and feel them out," said Sam. "Afterward, we can decide if we should ask them to join forces against the Fleas."

Hooker and Tom agreed, but the Captain had a little bit of a different take on what to do next. The Captain, who was far older than anyone realized, had been cloned many times before the ship even left Earth. He had been part of clandestine operations before and knew that the best way to enlist new people was to be open and honest with them up front.

"The people on this ship can almost always tell when a person was telling the truth," explained Volkov. "I will give them more information up front than was needed. Then, I'll decide immediately after talking to them both."

"Decide about what?" asked Hooker.

"You mean..." began Sam.

"If I decide against taking Whipple or the Chief Engineer into the fold, I will make them disappear forever as soon as possible," interrupted Volkov.

Suddenly, Sam realized just exactly how dangerous this whole thing was becoming. People were changing right before his eyes.

***

The Captain called the Chief Engineer and the First Mate to the bridge. He met them in his ready room located just off the bridge.

"Gentlemen, I want to know how long you needed to prepare the ship for the flip so that the ship could begin decelerating for the last half of the trip," he said.

Neither one of them could give a straight answer. They both explained that the ship's equipment needed routine maintenance; it had been mostly ignored for the last 300 years.

Then, the Captain had an idea.

"Both of you must immediately begin a thorough inspection of the ship's condition. Then come back to me with a plan and a schedule to accomplish the flip," he ordered.

Next, the captain called Sam, Hooker, and Tom for a meeting.

"I decided to only ask for a readiness report and a timetable at this point," the Captain said.

Sam, Hooker and Tom looked at him with puzzled expressions.

"Let me explain what I'm up to. Once the preparations begin, it will be in the different department's best interest to allow anyone involved in the flip full access to all the departments."

Hooker realized this would solve his problem. He now had an excuse to awaken as many new security and marines as he needed. Tom saw an even bigger opportunity. The flip was going to require that all the shuttles attached to the outside of the ship be controlled by one computer that would be located on the bridge. The preparations to accomplish this would give him and his men access and control over every computer aboard during the maneuver. Originally the plan was to get the shuttles ready before launch, but there had not been enough time.

Sam realized things were going in the right direction so far. "But security and Tom's computer engineers would need escorts when going about their duties," he warned.

Hooker agreed and, with the new cover, expected to have the necessary number of agents awake and ready for action soon. Tom decided not to make any substantial changes in the training of the men he had gathered from all parts of the ship. He wanted to return his agents as soon as possible to their respective departments and begin their spying activities. Their instructions were to collect any data they deemed useful and to remain undetected until he contacted them.

***

After some time, the Chief Engineer and First Mate contacted the Captain to report back.

Captain Volkov contacted the other members of the gang of four. "I am going to secretly record the meeting, rather than try to relay to you what Schmitz and Whipple said. Then if it turns out the two of them have to be eliminated, everyone will have the same information available. This will help us make the decision," he told them.

The meeting started off well enough. Both men saw the flip as doable without too much risk. They were not able to give a report on the ship's condition, having only accessed a dozen shuttles out of 1,800 minus the 225 shuttles destroyed when escaping Earth. The missing shuttles presented a problem.

"Many of the empty shuttle locations need a shuttle located there to provide maneuvering thrust. So, we will have to move shuttles," said Schmitz. "This will require pilots, but they are all in cloning storage. This will slow things down further."

"When the shuttles start the turn to reverse the thrust 180 degrees, the computer simulation predicts that there will be a slight tilting of the ship, creating the effect like one would experience gravity as if standing on the side of a hill," reported Whipple.

"How much tilt?" asked the Captain.

"The computer hasn't been able to determine that," replied Whipple. "The whole ship is going to need everything that could move to be securely fastened down."

"Now I want to know from you if you have confidence in the crew members in your respective departments," the Captain asked.

Both men looked at the Captain. One had a surprised look on his face; the other did not.

The Captain went on to explain what he meant. Neither man acted as if they understood how much the Fleas had infiltrated the various areas of the ship. But both agreed that they needed to be stopped.

At this point, the captain silently signaled for security to enter the room. The First Mate, to his surprise, was placed under arrest and removed from the room. His faked surprise reaction gave him away as a Flea. The Chief Engineer just sat there with a puzzled look on his face in silence.

A few moments later, the secret observers entered the room. Now, the Chief Engineer really looked worried. They all took a seat, and the Captain began to explain their plans.

CHAPTER 11

Sam's Dilemma

They were all going to have to work together. Their first decision was what to do about the First Mate. The position of First Mate was crucial to the success of their plan. After talking it over, they all agreed not to eliminate him yet but rather to re-clone him and remove the Flea syndrome. Hopefully, this would solve the problem, and they could bring him back into the fold.

"Well, during re-cloning, it is possible to achieve much more. The subject's behavior can be changed drastically. So, beliefs, likes and dislikes can all be changed to suit whoever oversees the cloning," Hooker explained. "I'm 100 percent sure I can return the First Mate to us and that he would be a perfect ally in the coming battles ahead."

This added information was going to be a game-changer.

"Since lift-off, the cloning department has been working on implementing the theory into practice. My technicians had been involved in cloning research since the earliest days. Before the voyage, they always suffered from a lack of subjects for clinical trials, but the crew on board presented all the subjects they would ever need for experimentation," Hooker stated. "The good news is that I don't think the science department and Dr. Harris have the knowledge to accomplish the same severe behavioral changes my techs have managed to accomplish."

"Why is that?" asked Sam.

Hooker chuckled. "My techs have disregarded any patient safety protocols that interfered with speedy results, and they have destroyed hundreds of lives," he said. "The workers that found themselves aboard the ship when it launched and were unable to return to Earth had been put in storage. Many were awakened as promised and then used to further clone research."

The meeting broke up with the following plan of action. Clone the First Mate again. When that was done, they would all talk again. Everyone needed to make plans.

"I will inform you once the First Mate is ready to rejoin us," Colonel Hooker stated.

***

Sam had safely returned to his quarters and was securely locked down. He began to go over in his mind how the crew had changed over the last 200-plus years. But he could not exactly put his finger on it.

Sam was one of the rarest of all savants. He looked and acted completely normal to everyone else, except he was not normal. He could, if given a little time, remember every day of his life in great detail, along with everyone he dealt with on any given day. When he realized he had this ability, it seemed like a curse to him in his early years, and he spent his entire life hiding it. Everyone who knew about it was long gone. Sam had not allowed anyone with knowledge of his abilities to join the crew.

He soon realized that what was changing his crew was changing him also and he realized he was uniquely qualified to find the underlying cause of the mystery. The physical environment on the ship was not any different from what people on Earth experienced when on long deployments to remote locations.

Then, he had an epiphany. The one experience that everyone shared aboard the ship *New Hope* was CLONING. Everyone awake during the journey so far had been cloned five to seven times!

Sam remembered his first cloning. For safety reasons, and in secret, he had given himself every physical scan and mental workup possible for a baseline to compare when he woke up. He had been worried about the Flea effect.

Sam decided he would work up a spreadsheet on his physical and mental condition before the first cloning on the ship. He had had one cloning on Earth before launch. Sam began to assemble his data. The physical data was the easiest to put together. As they came together, he did not see anything unusual. He, like everybody else, had noticed that with each cloning, a person's physical health problems that they inherited began to diminish. This had been one of the original reasons for the big push to put cloning on the fast track. Multiple cloning had never been intended, nor tested, on large groups of people on Earth. Only now was that happening.

Sam set to work. He gave himself every test he and the ship's computer could provide. The computer soon produced a voluminous report. Sam almost did not believe the computer's conclusions. His physical condition was now off the charts compared to his physical condition when the voyage started. However, mentally, he had changed for the worse.

His first mental work up had shown him to be an extremely well-adjusted individual in all areas tested. Now he realized the tests indicated he could be a monster if he needed or wanted to be.

He immediately realized this meant that every crew member was now exactly like him, completely untrustworthy.

Sam wanted the thrust reversal to be rescheduled to gain him time to make whatever changes necessary to regain control of the ship. He admitted to himself that it may not be possible. Sam had no idea who he could trust; he only knew he was going to have to have help. Hopefully, he will be able to accomplish this without explaining it all to his coconspirators.

He decided to concentrate on exactly what and how these changes were occurring. Each cloning produced subtle changes. The physical changes were accomplished in most cases on purpose – to make the crew members better suited to perform their duties aboard ship. The physical changes dealt with rearranging a person's DNA to obtain the desired results. So, you were not exactly cloning or making an exact copy of an existing human body. The DNA and all the data involved were sequenced to produce a desired body type. So, this allowed the body to improve with each procedure.

Sam sat down in his chair for a long, long time, thinking about what he had just worked out in his mind. This explained only part of what was happening. A person's consciousness was removed, and subtle changes could be made to their memory, but nothing was usually done to change their personality or behavior. The mind, except for additional memories, was an exact copy.

Sam sat completely still for a long time as he realized he had probably stumbled on the answer. What happens when you make a copy from the last copy over and over? You get a fuzzy-minded human copy.

This had to be the answer as to why the ship's crew, to put it bluntly, turned into a ship crewed by psychopaths.

Sam realized this was the reason none of the plans the gang of four planned ever seemed to be carried out. Investigating the ship's middle section, awakening additional security agents, and the problem with the First Mate – all this would never happen and would soon be forgotten.

Sam realized things had not gone at all as planned. *The only thing I have accomplished is to let an environment develop aboard the ship that guaranteed failure and the end of the human race*, he thought.

His new plan was to stop having any planned meetings and not share information with anyone. He was going to operate on a one-on-one basis. Over the years, Sam discovered that people never said the same thing twice. There were many reasons for this, but the main reason was they just could not remember. On the other hand, he always remembered and had been outsmarted only a few times in his life. Many a rival had disappeared and gone quietly into the night.

He smiled when he thought of how many people he had killed. Suddenly, his smile disappeared.

This is how a psychopath must feel, he thought, alarmed.

He only hoped he was up to the task of getting the ship back in order. He had 200-plus years to get the job done. Sam was probably the only man aboard who realized that the length of the voyage had been only a guess.

Brock, the Board Chairman, also knew this. Not a single clue had surfaced about his whereabouts or if he was still alive.

Sam called Tom. "Go through the complete list of everyone who is in storage. See if you can find Brock."

A few minutes later, Tom called back.

"Brock is not on the list," he reported.

CHAPTER 12

New Alliance

Sam realized the cloning of the First Mate could prove to be a trial run towards making crew members into trusted co-conspirators. He went through the records of the techs working in Colonel Hooker's cloning department. He found most had been cloned once and the next largest group twice. Their preflight mental tests revealed they had been extremely well-adjusted. He decided to approach techs who have had only one cloning.

Sam selected two techs at random and called them to his office. He began by asking them both how they had managed to only have one cloning, since the ship departed Earth almost 300 years ago.

Their explanation was they had planned to stay in storage the entire trip. Colonel Hooker had never given them a reasonable explanation for waking them up. They both added an unusual fact. The Colonel was slowly swapping the techs in his department with crew members who had only been cloned once, if possible and at the most twice. Neither tech offered a reason nor seemed to think much about what the Colonel was trying to accomplish, but Sam did.

Sam quickly realized that maybe he was not as smart as he thought he was. He was going to have to figure out a way to approach the Colonel. He decided to be straightforward and approach the colonel, and he would be armed when they met.

He contacted the Colonel and asked for an appointment, just the two of them. The Colonel replied by giving him specific instructions on where to meet. He had an eerie feeling the Colonel had been expecting his call. They met in the middle of the ship, where no one really knew what was going on. The cameras were still on, but large areas were still unseen by them.

Sam arrived by the secret passageway. When he stepped out of the hidden door, he was startled to find two armed guards, both wearing the ship's security uniforms.

"Are you Sam Otto Bernard?" one of the guards asked politely.

"Yes, I am," Sam replied.

"Follow us, please."

They began what turned out to be a 30-minute walk through a heavily wooded area. Visibility was down to about six to eight feet in any direction, including up.

This explains why nothing ever showed up on the ship's security cameras, Sam thought.

He almost fainted when he suddenly stepped out into a large, undeveloped area. There were buildings that looked like barracks for a couple of hundred soldiers. Sam stopped and looked in every direction. His guides stopped and stepped back about ten feet from Sam and were watching him intensely.

Suddenly, the Colonel stepped out of what looked like the headquarters building and stood on the porch. He waved at Sam to approach. Sam started to walk to the building slowly.

"Stop," the Colonel ordered Sam when he reached the first of four steps. "Do we need to disarm you now before we allow you into our headquarters?"

Sam realized his life depended on what he said or did next.

"I will gladly turn over my weapon," he said, hoping he sounded convincing.

He reached into his jacket and handed over his weapon. The Colonel accepted the gun, inspected it, removed the bullets, and returned it to Sam.

"When and if you leave this place, the bullets will be returned."

They entered the headquarters building and walked down a passage. Finally, they reached an office with a sign on the door: Office of Colonel Hooker.

Sam was completely taken aback by what he saw. He waited quietly for the Colonel to begin.

"Alright, Sam, forget or at least put into the back of your mind what you think you know. Board Chairman Brock is alive. He had been in storage since almost the beginning of the voyage. He was cloned about 175 years ago," the Colonel explained.

Sam heard a movement behind him. He turned around. It was Brock.

The Colonel and Brock stared at Sam intently as if waiting for him to start the conversation.

It took Sam several minutes to recover from the shock of seeing Chairman Brock. *How had they managed to hide Brock all these years? Dozens of people had to be involved*, Sam thought.

At first, Sam wanted to begin with softball questions but then decided against it.

"What do you plan to do to get rid of the Fleas?" Sam asked.

"We are going to get rid of them one way or another."

"But Colonel Hooker, over the last 300 years, you and I have made all kinds of action plans, but nothing has happened to complete any of them. Why?"

"The chairman and I were trying to buy time to come up with a solution to the cloning problem," Colonel Hooker replied. "We always gave the appearance of being helpful, but truthfully, we slow-walked or stopped everything you tried to accomplish."

Sam felt a shiver running down his back. "Then what made you approach me?"

"When you searched our personnel records and asked the techs to meet with you, we became suspicious. We had the techs drop a hint about the workforce comprising crew members who had been cloned only a few times. When you called, we knew you had figured it out."

Sam tried to swallow, but his throat felt dry. "Who besides the three of us know about the cloning process?" he asked.

"Only the techs performing the process and some researchers working on the problem," Brock replied.

"Have you made any progress to correct or at least arrest the damage to a person's mind?"

"Not much," the Colonel answered, not sounding sincere.

"Colonel Hooker, I have not been able to research all available information regarding a crew member's mental health. Can I please have access to these records?" Sam asked.

"Yes, soon," Hooker replied.

Sam realized then they were not being entirely honest with him. They had hidden information. His searches had been stopped in such a way that he had not figured out how it had happened until this very moment.

Whatever your plans are, if I want to stay alive, I will have to join you and appear to help you accomplish your goal, Sam thought.

Sam, at this point, was at a loss for words. Without knowing their plans, he knew he was powerless.

The ship's crew was hopelessly divided. Sam felt completely alone. Any crew member who was awake now had changed and was not the same person who had started the voyage. The crew was being cloned into two vast armies of fanatics. The ship did not have a chance in hell of surviving another 300 years if a third group was not organized to stop them.

Sam gradually got control of his emotions. The Colonel noticed it.

"Look, Sam, I don't know where to start. So, it might be easier for you to understand if I start today and then work back in time," he said.

Then, the five remaining board members entered the briefing room and took seats at the table. Sam knew Brock very well from the early startup days back on Earth, but these faces were not familiar to him. Everyone in the room sat patiently, waiting for the Colonel to begin again.

"Alright, let's start. The base camp in the ship's middle section had been planned since the beginning, along with the location of the security cameras. This had worked out better than we had hoped.

"I am still waking up additional security personnel, and they are being housed and trained in Section 3, the largest section on the ship. It is twenty miles in diameter, and the ceiling is 800 feet high. There are over 150 decks of various heights and hundreds of square miles to hide our activities."

Out of the corner of his eye, Sam noticed six security guards quietly entering the briefing room. He quickly realized they all shared the same physical characteristics. It was barely noticeable, but it was there.

Sam managed to get his mind back to the matter at hand: his survival. He realized the Colonel had waited until all the players were in attendance to begin.

Since day one of the Project, the Colonel, Brock, the board, and Dmitri had been cooperating to gain complete operational control of the ship. They had built the middle deck complex secretly before launch and completed it after takeoff. The complex garrison comprised approximately 300 men. They were in training, and they were planning to take control of the ship when all the preparations had been completed.

"The Food Production Department, headed up by Anastasia, had successfully been taken over within the first 100 years," said the Colonel. "The Engineering Department of Wolfgang Schmitz, ship's First Mate Oliver Whipple, Medicine and Science Director Dr. Emily Harris, and ship's Maintenance Chief Qiao Hui... These departments are now in various stages of being taken over by the Fleas."

Sam, upon hearing this startling revelation, lost his poker face. It was apparent to everyone in the room that he was completely gobsmacked. He had a look of complete and utter terror on his face.

"You see, my group is actually the Fleas," the Colonel explained. "This enabled us to conduct clandestine operations in all the other departments with the cooperation of the department heads. They were, in effect, looking for themselves. This turned out to be the perfect cover and allowed us complete freedom to access all the data in the uncontrolled sections of the ship."

Sam soon recovered enough to ask questions of his own. "When did you discover the effects that repeated cloning was having on the ship's crew?"

"Right after launch we began to continue research on the human mind, and we cloned hundreds of people. Doing computer searches looking for data on subjects who had been cloned more than once, by chance, we found the data you had assembled," said Colonel Hooker. "We realized there have been remarkable changes in your physical make-up. So, we decided to watch your actions and see if the changes made you into a candidate for joining our little group.

"You have spent the last couple of hundred years looking for Fleas. We think you want what is best for the remaining members of the human race. But your new personality demands that you want to belong to the winning side."

Sam realized they were right. He no longer seemed to have a conscience. Survival seemed paramount, and only survival mattered.

"Very well then. I will share all the information I have with you," Sam stated.

He began to have doubts about their sincerity. He knew he could be easily replaced. The bullshit story about discovering his cloning data was the reason he was still alive, which made no sense at all. Sam knew he was still not breathing because of their goodwill; it had to be something else.

"Well, you will have to... I will need some guidance in the future when dealing with the crew," said Sam. "Is Thomas Hacker in this group?"

"Yes. Everyone is aware except the Chief Engineer, the Maintenance Director, and Dr. Emily Harris."

"Alright, then. Brief me on the takeover plan for the rest of the ship," Sam asked.

"It will happen after the flip. As you know, this is planned to happen as soon as possible. The lower part of the middle section containing farmland and thick forest is where the security complex is located. The area above Section 3 is the security department. The First Mate and the general crew are directly below the security section. Dr. Harris is next, directly below the First Mate and directly above the farm and forest department. When the time comes to take over, it will be a simple matter to close off the two subsections occupied by Dr. Harris and the First Mate."

Sam felt like they had been truthful concerning the coming takeover. The answer to the next question was going to be revealing.

"So, what are you going to do with all the crew members who were not indoctrinated with the Flea syndrome," Sam asked.

"Any who were killed will be gone forever. There is nothing I can do about that," the Colonel said curtly. "Survivors who are critical to the operation of the ship will be allowed to continue doing their duties. Any other survivors will be placed in the cloning storage bank for future awakening and use if needed."

"How much new knowledge and abilities have been discovered about the cloning process?" Sam asked. "Who possesses this knowledge?"

"I have control of all cloning on the ship, except for Dr. Harris, who is under my control at my laboratory," the Colonel confessed.

For the first time, Sam detected a hint of plain old bragging in the Colonel's voice.

"What are these breakthroughs?" Sam asked. "And who is in charge of their use?"

"I make all the important decisions," the Colonel replied, again with a certain hint of self-importance.

Sam again realized he was the only one of the non-security people who was armed but he had no bullets. *No wonder he made all the decisions*, Sam thought.

He looked at the other people in the room. *There are uneasy people in this room, and I'm one of them*, he thought.

"Colonel, can I talk with the cloning scientists?" Sam asked. He still had not gotten a straight answer about cloning improvements.

"Yes, I will set up a time for you soon," the Colonel replied.

"Can I also be taken on a tour of the complex?" Sam asked again.

"Certainly. However, the tour will require several days, so you have to stay at the complex."

Sam followed a sergeant to his temporary quarters. Someone would come for him in a few hours to take him to the mess hall.

CHAPTER 13

The Sergeant

Strangely, it just now dawned on him that nothing had been mentioned about the Black gang in the engine room and Wolfgang Schmitz, the Chief Engineer.

The engine room crew had the smallest number of personnel of all the departments aboard the ship. Keeping the engineers functioning was paramount for the ship to complete the voyage with one G of gravity. Sam could think of no reason they had not mentioned the Black gang. What could possibly be the reason? But Sam decided when they came to get him for the next meal, he would keep quiet about the omission.

The same Sergeant came for him shortly. He obediently followed him to the mess hall and was immediately astounded by the size of the room and the hundreds of crew members. He had only seen a small number of the assembled crew members before. Sam was led to the head table and showed a seat.

The Colonel introduced Sam to the men and women who were already seated.

"Sam had been the Project's Chief of Operations since the very beginning. Early on he had been responsible for everything that had allowed the mission to reach the halfway point," the Colonel explained. "But now, the responsibility to bring the second half of the trip to a successful ending is going to fall on everyone in this room. You all have to put the mission first."

After the meal, Sam found himself again under the watchful eye of the Sergeant.

"So, what do you want to see?" asked the Sergeant.

"Why don't we start a walking tour? I will ask questions as we go," Sam replied.

They began to pass buildings and shops and walked for about six hours. Then, they returned to the barracks where Sam was to be quartered. Sam was puzzled he had not seen enough infrastructure to house or train the number of crew man that were in the mess hall earlier. Mostly, they had walked down narrow paths under a canopy of trees to avoid the cameras.

Sam decided he would approach the Sergeant with this observation and see what kind of reaction he got.

"Where is the rest of the secret base? What you have shown me could not possibly be the entire campus."

The Sergeant came to an abrupt halt and put his hand on his sidearm as he turned to face Sam head-on. He seemed to be uncertain about what to do next. Then, he asked Sam a strange question.

"In the closing weeks before launch, you interviewed dozens of security people to fill out the ship's security needs. You interviewed me. You accepted me and dropped someone who had been previously accepted to make room for me. Do you remember me?"

Sam had remembered him but did not acknowledge it. He was the only candidate who brought his family to the interview. This impressed Sam. He could not let the Sergeant know that he remembered it at all.

"No, I don't think so. But tell me about the interview, and maybe then I will remember," Sam replied.

"I brought my family along to try and get the idea across that there would be four lives impacted, not just one."

Sam remembered that men who had families had, as a last resort, agreed to leave their families behind for a spot. This was exactly the kind of person they needed, but for some reason, Sam decided to give the four of them a spot. Then, he had not given it a second thought in over 300 years.

This Sergeant just might be my first recruit, Sam thought.

"How many times have you been cloned? Where is your family presently?" he asked.

The Sergeant's answer was what he hoped it would be. He had been cloned only once. He had been awake for approximately 25 years, and his family was still in storage. Sam realized that his personality had probably not changed much, but the Flea syndrome would be present. He could deal with that when the time came if he had to.

Sam began to go over the Sergeant's record in his mind. The Sergeant's name was Toby Sawyer. When interviewed, he was 55 years old. His wife, Janice, was 53, and neither of them had been cloned yet. Their only daughter was Alice, aged 30. She had advanced degrees in medical records and computer science. As it turned out, she had been instrumental in getting all the world's databases downloaded to the ship's computers. She had reduced duplication to almost zero thereby freeing up enough room to store all the data available.

Their only son, David, was 35 years old. He had been a college professor with several doctoral degrees in religious studies concerning the world's religions, past and present. Sam immediately realized David might be the only person on the ship with a working knowledge of the world's religions. He was going to be a valuable asset, he hoped.

"I need time to contact my superiors," Sergeant Sawyer said. "I need to discuss this latest development with them."

He disappeared for a few minutes. When he returned, he had a smile on his face.

"The Colonel's orders are to show you anything you want to see and to answer your questions as honestly as possible."

CHAPTER 14

Little Known Areas

"Where is the base located?" Sam asked.

"Follow me."

They soon came to a heavily wooded area remarkably close to the ship's outer hull. Sam entered closely behind the Sergeant. The Sergeant stopped abruptly before what looked like camouflaged elevator doors. He pushed a hidden button, and they waited.

Shortly, the elevator arrived, and the doors opened. They waited for a dozen armed security personnel to exit. They all greeted the Sergeant by name.

Sam wondered whether the elevator was going up or down. He went over the ship's blueprints in his mind and did not remember any elevators in the middle decks. The elevator went down to the next level. The doors opened.

Sam was visibly shaken by what he saw. At this deck level, the ship was approximately twenty miles in diameter. He had an almost unobstructed view of the whole area. He could almost not believe his eyes. The bulkhead looked like it was at least 800 feet high. There were even ultralight planes overhead, hundreds of buildings and men. They could only be described as soldiers.

How had they gotten this secret deck past me during the ship's construction and for hundreds of years? Sam thought. *I wonder what else is going on in the areas of the ship not under the Colonel's control.*

He truly did not know his next move.

The Sergeant was looking at him intensely. "Follow me," he said softly.

Sam had only taken a dozen or so steps away from the elevator doors when he became dizzy. The immense size of the hidden deck was more than his senses could take. He sat down on a log to try and gather himself. He was having a panic attack. Almost 300 years in the confines of the ship had not prepared him for suddenly seeing approximately 300 or so square miles without any warning. He was at a loss for words and beginning to experience difficulty breathing.

Sergeant Sawyer noticed Sam was not next to him. He turned and saw Sam sitting down. He smiled to himself. He had seen this reaction before from people who visited the middle deck for the first time. He returned to Sam's side and sat down next to him. He patiently waited for Sam to gain control of himself.

"How had this been done right under my nose?" Sam asked hoarsely.

"They used one of the oldest tricks in the book," the Sergeant explained slowly. "No one had a complete set of the ship's blueprints, nor was a complete set ever allowed to be in one place together.

"Who knew?" asked Sam. "Who is responsible?"

Sam hoped he had not gone too far, but this last question might reveal their intentions as to his future.

"Chairman Brock is responsible," replied the Sergeant. "He was actually the one in charge and had been running the ship since the launch."

Sam now realized Brock's disappearance in the beginning was done to allow him to orchestra the operation of the ship in secret.

All the information about Chairman Brock since its launch had been misleading, pure and simple.

"Do you feel well enough to continue our tour?" Sergeant Sawyer asked.

"Yes, please!" Sam exclaimed.

They soon approached a crossing in the road. A detail of four armed soldiers was waiting for them and they fell in behind them. Sam did not like this but decided to hold his tongue.

"When is the ship scheduled to flip?" asked Sam, trying to change the subject.

"I don't know, but my best guess is very soon. The preparations are almost completed," the Sergeant replied. "The Colonel ordered that we guard you until the flip was completed. We've prepared a room in the officers' quarters for you."

Sam felt like he had been duped. He came here looking for more information and he now realized he had been marginalized, and his movements restricted.

By now, Sam had completely recovered from his panic attack and asked questions about everything he could think of. They passed dozens and dozens of structures, weapon firing ranges, and everything else one might think of that would be needed to take control of the ship. One of the buildings looked like a prison.

"Is that what I think it is, Sergeant?" asked Sam.

"Yes, if you think it is a prison."

Sam was not sure if the Sergeant was aware of the evolution of the crew slowly turning into psychopaths.

"What brought about the need for a brig that large when you could easily just put them in storage until needed?"

The Sergeant hesitated ever so slightly before answering.

"Are you aware of the fuzzy copy problem?"

"Yes, I am."

"The fuzzy effect gets worse exponentially with each cloning. There was the case in the engine room; because of the high radiation count, the men had to be cloned every 15 to 20 years, meaning some crew members had been cloned 25 times. This proved to be a disaster. There are men in the brig who do not behave in any way like human beings. There is a large research effort underway using them as guinea pigs to find the solution. Crime and cruelty throughout the ship are increasing daily. The engine room is completely isolated from the ship. The Chief Engineer had not been seen or heard from in years."

Sam took a deep breath. He felt overwhelmed.

"Why are there so many men here on the hidden deck? Why do you need so many men? There seems to be more security people here as were awake in all the other sections of the ship combined."

Sergeant Sawyer offered no answer and remained silent.

They walked on, but by now, they were both getting tired and hungry. They came up to a temporary mess hall and decided to grab coffee and a meal. Throughout the meal, Sam continued to be amazed to hear what the Sergeant was sharing.

The engine room had, in the recent past, been vented to open space. This killed everyone except the Chief Engineer. He had been taken prisoner earlier, and he was cloned.

"The radiation problem in the engine room became a big problem, "the Sergeant explained. "After the venting, the Captain and the Colonel decided to act. The Colonel used the secret passageways from the bridge to the engine room and sent a group of computer techs with a security escort to make changes in the ship's main engine controls. These changes made it unnecessary for a living crew to be present."

"Surely the crew had noticed?" interrupted Sam.

"The crew members who were killed when the engine room was vented were replaced by a temporary crew. When the changes were completed, a full-time live crew would not be needed. The technicians were alarmed when their security team killed twelve of the temporary crew who resisted. The temporary crew did not react well to the situational pressure and the radiation, but the crew members who cooperated all returned safely to their regular assignments."

"No one ever felt using only remote controls was the best solution. The ship had, for years, used specialty robots to perform tasks outside the ship. The robot designers set to work developing robots to take control of the engine room. The job of putting the robots in place is well along."

"Is this what is holding up the flip maneuver?" Sam asked.

"No," the Sergeant answered. "The ship is going to be returned to its construction condition. This will allow it to survive weightlessness. That translates to its condition before launch when it was parked on the dark side of the moon.

"To survive the flip and begin the deceleration was going to require the cooperation of everybody on board. So far, it looked to the Captain and the Colonel that this was going to be an impossible task. All the department heads had the information necessary to make it happen, but so far, truly little has been accomplished."

Sam realized that all the various department heads were trying to protect their turf. The department heads had never had to cooperate on this magnitude before and simply did not feel they had to know. They did not know one another. They all realized what terrible things they had done themselves to remain in control, so they knew they could not trust anyone. This distrust carried over into the daily operation of the entire ship.

Sam had a bad feeling about why he had been detained and allowed to live this long. He decided to just try and remain a bit of a player and wait for the Captain and the Colonel to approach him. He would find out soon enough if his suspicions were correct.

He did not share his thoughts with the Sergeant. The Sergeant seemed to have grown comfortable talking to him, so Sam decided to try and ask some innocent questions.

"Sergeant, would you like to wake your wife or children? Perhaps I could be of some assistance."

"Not yet," he answered without changing his expression one bit. "All my family would have to be awakened by Dr. Harris."

Things were going just as he had hoped. Sam's next question was a dangerous one for him to ask.

"Sergeant, are Dr. Harris's laboratory results any different than the Colonel's?"

"Yes, they are."

"Can you give me an example?" Sam asked.

The Sergeant took a deep breath before answering. "Each side is changing newly cloned people in diverse ways. Dr. Harris is all about (for lack of a better word) making people from scratch. Changing whatever is necessary in their physics to make a completely loyal crew member. They are programmed to, if necessary, die for her. They all look alike physically. Most of their memories about Earth are removed, especially family. This has allowed Dr. Harris to create a group of stone-cold killers.

"Dr. Harris has also begun the process of awakening more women than are necessary to replenish the crew. These women have had all their memories erased about their time before being put into storage. Dr. Harris has also started cloning people, men and women, in pairs. They are designed to be inseparable and almost think as one person. They are also sterile."

"Why do you not share the same physical characteristic traits as the other security people?" Sam asked.

The sergeant's answer floored him.

"I was awakened recently for the first time. The 25 years had been a lie. They knew I had met you before the launch. They hoped you might remember me, trust me, and believe what I tell you. So, they decided to not make any changes in the way I look."

Sam took a few moments to digest this.

"Sam, do you recognize me from the meetings we had before launch?"

"Yes, you may look the same, but you are not the same. You are the only security person I interviewed that I remember," Sam replied. "Things stick in your memory, and some do not, and for no apparent reason. But there is a reason I remember you, and an incredibly good one. You were the only applicant out of thousands who said you wanted to be a part of this ship's crew, but only if your family could accompany you. That has stuck in my memory all these years. I do not remember your wife's name or the names or the number of children you have, if any."

This seemed to satisfy the Sergeant. Sam realized this man was his intellectual equal, but without his memory. He was not the same simple family man he had interviewed 300 years ago. But Sam knew he was, and he had just missed it.

Sam was really worried now. Did they know about him having hyperthymestic syndrome? He was an extremely rare Savant; Sam had a perfect recall of everything he had ever been exposed to. When he had his first cloning back on Earth the techs had managed to alter his looks and behavior in such a way as to make it impossible to detect his condition. The techs had been aware of his secret. They had all been sworn to secrecy. His secret had been kept for over 300 years. Sam was the only person alive who had been at his cloning. The technicians involved had all disappeared over 300 years ago.

"Let's return to your quarters," the Sergeant said.

Soon, they were on their way and arrived at Sam's quarters.

"I will pick you up after you get some sleep," the Sergeant said as he was leaving.

How would anyone know how much I sleep if they were *not watching me right now*, Sam thought.

He hoped he would get the answers tomorrow, but first, he needed to gain access to the ship's computers. There was a computer on his desk. No way was it working, or it had to be bugged. There was only one way to find out. He turned it on and found everything working normally.

The first thing he did was get a movie up and running. The movie would shield his activities on the dark web.

CHAPTER 15

The dark web and the ship flip

Sam was sitting all alone in the room provided for him, just as he had 300 years ago. He was feeling the same anxiety he experienced then. The unknown future that lay ahead for him and the entire ship had petrified him. Everywhere he turned, people were becoming increasingly savage and unpredictable. He was losing hope.

He turned to the computer and began to think about how he was going to wake up the dark web. He sat there for a long time, making sure he remembered the procedure.

The first passcode was 99 digits long, but only every other digit made up the code, and the other random digits were just fillers. When the preceding process was successful, it would set up the need for a second passcode. This one had 100 digits.

Only two people knew about the dark web, and the other person was in storage. Alice Sawyer must be kept in storage until he was ready for her help. His original plan had been to have her awake when he was undergoing cloning to protect him, and he would do the same for her.

Alice had been put in storage from the start. She was considered a nonessential crew member. There had been no way for him to stop it. He hoped to change that soon.

Sam remembered the two codes perfectly. He began typing them in just as smoothly as if he were reading from a list. Every one of the ship's servers had only a few GBs of its storage dedicated to the dark web. The dark web software had successfully managed to remain in the background undetected, and after activation he prayed it would remain so.

He chuckled. He had not used that term since he was a child. The voyage was changing Sam every day and probably also everyone on the ship.

Every crew member had a serial number. The dark web could locate anyone on the ship if you knew their serial number. Surveillance was then available. Sam began going over a list in his mind of people to spy on.

He found the Board Chairman Brock and the Captain together on the bridge. Bingo. They were in the final planning stages for the flip. This was exactly what he needed to know. It was scheduled to start in 12 hours.

Sam's interest was piqued when First Mate Oliver Whipple joined them.

"All areas of the ship are ready," he reported. "The whole ship had been returned to its pre-launch condition."

The men discussed several areas that were of concern.

"Just remember, neither I nor my men inspected the maintenance area, food production, and most important of all, Dr. Harris's medicine and science department," said Whipple. "There are no crew members in the engine room, only robots. A few hours before the flip, crew members will enter and oversee the robots' shut down and restart the ship's main engine."

The group was extremely interested in what Whipple had to say about the shuttle preparations.

"What took so long to get the shuttles ready to play their part in turning the ship 180 degrees so as to allow the ship to decelerate at 1 G for the last half of the voyage?" Brock asked.

As if right on cue, Tom Hacker entered the room.

"Here you are!" exclaimed Whipple. "The one person who understands the most about the delay is Hacker. Tell us what took so long, Tom."

"Simply put, we had a lot to do. The shuttles, 225 to be exact, had been reprogrammed to serve as self-guided makeshift missiles. We used the shuttles that were attached to the ship in the best location to allow access for reprogramming and loading the hydrogen bombs. We had to replace half the missing shuttles by relocating others to get the proper thrust points to flip the ship safely over," Tom explained. "And that's not all. The crew members with the proper training had to be cloned. Because this was done in Colonel Hooker's limited laboratory, it had taken extra time, measured in years. A dozen or so crewmen had lost their lives."

The Captain got on the horn to issue orders to the entire ship. "Everyone not having duties involving the flip will be confined to their quarters. The clock is running minus 8 hours to the start of the second most critical maneuver the ship would ever perform."

The other men in the room had moved out of the camera view. The captain appeared all alone by choice.

At this point, Sam decided to find Dr. Harris. This proved easy to accomplish. She was in her conference room, along with Anastasia and Qiao Hui. They were agitated and all talking at once. Harris banged on the table for silence.

"I want to hear you – one at a time," she exclaimed.

Anastasia went first. "My department is in its pre-launch condition. I expect no problems. However, I had to allow crew members from other parts of the ship to have access to my department. I have done this against my better judgment."

The crew members were getting the shuttles ready to flip the ship. They had to exit the ship and fly the shuttles to the vacant docking locations on the ship's outer hull. Six pilots and technicians were lost in accidents. One was kidnapped by Anastasia's men, and the detail she was working with were all led to believe her disappearance was just another accident. This unplanned act done spontaneously turned out to be of immense value. Anastasia was terribly angry at her people in the field until she realized just how important the female tech turned out to be.

She had been interrogated and then put in storage. The woman was a computer specialist and had spent the entire voyage assigned to the bridge. They were able to retrieve memories from her unconscious mind she had forgotten. They now had almost a day-to-day log of everything that took place on the bridge. But the most important thing they now had in their possession was the security codes.

"We only have minutes before the engine will be stopped, and the ship will be weightless," Dr. Harris announced.

The operations order stated that upon the shutdown, the engineering department would pause for one hour and then proceed with the flip. This would allow the ship's personnel to deal with potential problems.

No sooner had the captain finished his announcement than the ship's engine shut down. It had been years since the crew had experienced the absence of gravity. There were all kinds of reactions, from ho-hum to panic. The varied reactions had been expected. The medical unit had made available a large number of tranquilizers. This proved to be a wise, if not suspicious, decision by Dr. Harris.

The Captain spoke again. "We have started the maneuver using the shuttles. Hopefully, we will achieve a perfect 180-degree reversal on the first try. Because the shuttles are attached to the exterior of the ship in a fixed position, we can only rotate the ship in one direction. We will know if we are successful in approximately eight to ten hours. Please stay calm; the flip will be over before you know it." The Captain signed off.

Sam decided this might be a suitable time to have another look at Dr. Harris using the dark web. Something did not seem right about Dr. Harris's disbursing drugs ship-wide and giving the impression it was a last-minute thing.

He found Harris and her staff still in the conference room. Sam relaxed and began to observe and listen very carefully. Apparently, he had missed the first few minutes of the conversation. They appeared to be about to break up when a clear voice sounded.

"How long is it going to take for the drug to have its maximum effect on those who took it?"

This question sent Sam into a panic. He knew neither the captain nor anyone else had any knowledge about Dr. Harris's plan. What was he going to do with this situation? He could not come right out and warn everyone because he would have no way of explaining how he knew. He could not reveal the existence of the dark web for any reason.

A plan began to form in his mind. Maybe there was a way. The drug had to be distributed to all parts of the ship, except the areas controlled by Dr. Harris. Food distribution had to be the answer. The big question was who knew and what was going to be the effect on everyone aboard the ship. Hopefully, after it is identified, any effects can be reversed.

The battle for control of the ship had started, Sam thought. *Which side am I going to join?*

Based on his present situation, Sam thought he had better join the Captain's side. There were a few more hours to wait before the flip would be completed. He suddenly realized he was hungry. He would go over to the mess hall as soon as possible and try and get the Sergeant to meet him there.

The ship's crew was successful the first time; the alignment was perfect. It was going to take 48 to 72 hours for the ship to achieve 1 G of deceleration.

Sam exited from the dark web. He then contacted the Sergeant and asked him to join him in the mess when he could.

"I'll be there in 30 minutes," the Sergeant said.

As Sam was turning off his computer, he heard the lock on his door click. He walked over, and yes, the door was now unlocked.

***

Sam arrived first, took a table, and waited. He stared intently at the people in the room. Several of the food handlers seemed to be acting as if they had had too much alcohol. But since early on in the voyage, the use of alcohol had just about disappeared. There were too many other recreational drugs available. None of them harmed the user's body, and the duration of the effects and the intensity could be regulated.

When the Sergeant arrived, he was accompanied by two armed soldiers, who did not look happy to be there.

"Sergeant, please order the two soldiers to sit down so I can talk to you all," asked Sam.

The two men sat down. Sam began slowly, not sure how to proceed. He knew that he was taking a risk.

"Sergeant, I want you and your men to look closely at the kitchen help and tell me what you see."

The three men did what Sam asked. The realization as to what they were seeing gradually became apparent to them. After about 15 minutes, they all looked at Sam with a blank stare.

Sam realized the Sergeant had become very suspicious. He was going to have to produce a good reason as to what first alerted him.

"Let me explain," Sam said. "As Chief of Operations, I had, at one time or another, approved the operation of all the food distribution on the ship. I had assumed this mess hall operated the same. The kitchen staff, therefore, ate first. It would be an impossible coincidence for the entire kitchen staff to get drunk or take a recreational drug at the same time. When last did you three gentlemen have something to eat?"

"Maybe 18 hours," the Sergeant replied. The two escorts nodded.

"These people also had not eaten in 18 hours. Please take me to the Captain."

They were soon on their way. The Sergeant let Sam and the escorts get out of earshot. He immediately called the Captain and began to try and explain.

The Captain listened intently. "Sergeant, send the two armed soldiers back to the mess hall and, if necessary, use deadly force to stop the serving of food until further notice," he ordered.

The men were soon on their way. Sam and the Sergeant decided to try and get to the bridge as fast as possible. The ship had not yet reached its goal of one G of deceleration, so it was easy running. Neither worked up a sweat, and they soon arrived at the bridge.

They were all there and were looking at him strangely.

"When did you have the first clue that something was wrong?" The Captain asked.

"Not until I arrived at the mess hall and observed the kitchen staff's behavior," Sam replied. "I still did not know exactly what Harris's plan is and still do not."

Dr. Harris's distribution to the entire ship of a tranquilizer, Sam thought, was strange. No one had been given prior notice. She expected everyone to immediately suspect that it was a trap to disable all the crew she did not control.

"Tell me if you know, Captain. Did she supply the drug to her people?" Sam asked.

"I do not know."

"Have you had the food analyzed for drugs, Captain?"

"We are in the process as we speak. It is not turning out to be something that seems easy to identify. The lab will notify me as soon as they know what it is."

As time passed, it became apparent the food was neither going to kill anyone initially nor debilitate anyone for long. Still, no one knew the long-term effects of exposure. They desperately needed to know what they were dealing with.

The Captain slowly looked at his staff and said plainly. "I want everyone on this ship to understand how serious the situation is for us. I am ordering all of you to immediately shut down all our mess halls. Anyone who issues or receives this order and does not obey is to be summarily executed on the spot."

The captain looked at Colonel Hooker. "How many security men do you have who, in your judgment, are ready for combat?"

"I can put 300 men in the field for a brief time. For a longer campaign... that will depend on the supplies available," Hooker replied.

"Get them ready to go as soon as possible. As soon as we obtain Dr. Harris' location, she will be your target."

The Captain did not need to say it, but he did anyway.

"Gentlemen, it will be over for us unless we can obtain food that is safe to consume. We are going to all be starving very soon if we do not figure this out."

"Captain, can I accompany Sergeant Sawyer? I have been trained in the use of firearms and believe I may be of help to the Sergeant," Sam asked. He did not want to miss the end of Dr. Harris.

The Sergeant gathered his company of four six-man squads, each led by a corporal. Sam realized that the security department had been reorganized and now operated exactly like the military on Earth before launch. He didn't know how many soldiers would be awakened for the coming battles.

"We are going to take control of the computers located in Dr. Harris's laboratory," said the Sergeant. "Thomas Hacker will be joining the team before we go to the computer lab."

The soldiers were going over their equipment checklist and getting ready for action. Then, Thomas Hacker joined them, handing Sam a weapon and a combat kit. Sam now had the same equipment as everybody else.

Thomas took Sam aside and whispered, "I did not ask the Colonel for permission to issue you a weapon. If this is a problem for you, just leave it behind."

"No, no, no problem at all," assured Sam.

I wonder what Hooker's reaction would be if he even noticed, wondered Sam.

They formed up and entered the secret security passageways. The trip to Harris's lab took about 30 minutes. They did not encounter any other soldiers, and they soon arrived at the chosen door, which would serve as their entry point. There were multiple doors.

"We are going to secure almost all the entry points, and no one will be allowed in or out," Tom told Sam. "A much larger force will spread out into the interior of deck three. There are three subdecks. We are going to subdeck two, where the cloning labs are located."

They began to go into the interior and were met by a group of Hacker's computer techs, who fell in behind them. Everyone was worried; they had not met any resistance yet. They soon reached their goal and did not fire a shot.

They entered the first cloning room and found the technicians in the process of finishing a procedure. The technicians were completely surprised and seemed to be completely in the dark as to what was taking place.

"We have taken the cloning labs," reported the Sergeant.

"Good! Report to headquarters and bring Sam. Leave Hacker – he already had his orders."

On the way, they had a chance to talk alone. Neither one seemed to have a clue what was going on. There has been truly little resistance anywhere on the ship.

The trip back to headquarters took about 45 minutes. There were armed soldiers everywhere along the way. Upon entering the large briefing room, Sam saw Dr. Emily Harris, Anastasia and Qiao Hui. They were completely surrounded by armed guards and were seated across the table from Chairman Brock, Captain Volkov, Colonel Hooker, and First Mate Oliver Whipple. There was a vacant seat at the head of the table.

"Please take that seat, Sam," said the Sergeant.

Sam sat down. He wondered why the ship's Chief Engineer was absent, but he was not about to ask.

Chairman Brock, ever the consummate politician, spoke in his "public" voice.

"What happened? The ship is in its 350th year of a 600-hundred-year trip. Does anybody in this room think we will survive the voyage if we continue our present behavior?"

"There is enough blame to go around," interjected Dr. Harris. "I feel I have been shunned since the beginning. I have been forced to cut my department located in section three off from the rest of the ship."

Brock took a hard line. "Your behavior allowed you to operate in such a way that has the ship in grave danger. Dr. Harris, you are now standing at the brink of a mutiny, and you may be punished along with your lieutenants. We are aware of your secret group, the so-called Fleas. There can be only one reason for their existence; you plan to take control of the ship."

Dr. Harris had a very menacing tone in her voice.

"You, Chairman Brock, the Captain, Hooker, and the rest of the conspirators are the group who pose the greatest danger to the ship. When you start cloning genetically engineered security people and their mates, you have started down a road that will lead to disaster. Everyone now understands, because of the fuzzy-copy effect, what is happening right before our eyes. Crime and mental illness are on the rise. Through our own actions, we are quickly becoming a mirror image of what we left behind on Earth. Only now there is nowhere to seek refuge from ourselves."

Sam watched this exchange intently. He genuinely believed that neither side understood the other. The mere fact that each side was attending these negotiations revealed to Sam that neither side really knew the other's strengths and weaknesses. Why was he sitting at this table right between them? He decided it was time for him to make a statement. But before he could open his mouth, Brock spoke.

"Dr. Harris, will you agree to continue in private?"

"Yes," she replied.

"Clear the room now," said Brock in a very loud voice. "I mean everyone except Harris and Sam."

Dr. Harris ordered her people out as well. Sam nearly fell out of his chair. Thank goodness it was going to take a few minutes to get everyone out of the room. Sam needed a few moments to gather his thoughts. Sam did not trust either of the two people sitting next to him.

But I'll agree and help with whatever the two of them want. If I could get them to trust me or at least give me some amount of control, and I had secret access to the ship's dark web, I would eventually get the upper hand, Sam thought.

CHAPTER 16

Sam's New Duties

"I think it's fair to say we both know Sam well," said Brock. "He might be the one to function as the go between. Correct me if I am wrong, but Sam, to my knowledge, has not attempted to gain control or make policy in any way in our sectors."

"Yes, I agree," said Dr. Harris. "I have seen Sam only a few times since launch and that was in the beginning."

To their surprise Harris also admitted she knew truly little about Sam's activities during the voyage. But then the tone of her voice changed to very menacing.

"You and the Colonel have created an army of killers. You now know this was not necessary. My group does not have any armed men available who were created to fight. What is most disturbing is the fact that your men now are stationed throughout the ship unopposed and control everyone on board. Control was not our goal. We both know crime is widespread. The Ship never had any laws and no one to enforce laws or a court to try suspects."

Suddenly, Dr. Harris looked at Sam.

"You said you were the Chief of Operations. How was this overlooked?" she asked.

Sam was caught off guard and he needed a little bit of time to produce a plausible answer. Harris and the Colonel both looked at him intently.

"I now realize my planning contains errors," Sam said after a long silence. "There was not enough data available to guide me and my staff in planning for a 600-year voyage. We resorted to computer models and applied plain old-fashioned common sense. This has not turned out too well up to this point.

"The biggest mistake was the segregation of the crew by sections. This has bred unfamiliarity with each other and led to distrust and eventually violence. I think the enormous amount of time that passed was not factored in properly. When we designed the checkpoints, that proved to be a grave mistake, everyone felt isolated, and this only led to more distrust.

"When the ship was in the process of leaving Earth and it became obvious, we were going to have to kill hundreds of fellow humans in order for the ship to escape. The ship's officers and I never gave it a second thought. This behavior was a bad example, it set the crew's mood early on the voyage.

"Cloning had not turned out to have worked out the way it had been planned. A large number of people wanted to be in storage until arrival. Soon people wanted to be awake the entire journey. These people were petrified they would be reengineered if cloned. They have formed all kinds of secret groups. The hope is that each group will be able to prevent any DNA changes among its members. The biggest unforeseen problem has been the fuzzy-copy syndrome. It destroyed the engine room crew."

"Please tell me what happened?" interrupted Dr. Harris.

"The unplanned increased radiation was caused by a design flaw," replied Sam.

"Why was the flaw not discovered during the initial main engine testing?"

"Yes, it would have been, but the only live testing the main engine received was a low-level test done during launch," said Sam. "The plan was to leave orbit slowly and gradually run the engine up. It would have given time to hopefully find and correct any problems. But we had to forego all testing and go to 1 G acceleration almost immediately in order to outrun the pursuit ships from Earth."

"I did not know about the lack of pretesting," said Dr. Harris. "How are the robots working out as the engine room crew?"

What Sam did not tell Harris was that when engineering developed the engine room robots, they did not stop there. They took the opportunity to develop some extremely dangerous robots. They could conduct operations autonomously. When the situation required it, they would call in for clarification. Had it not been for his access to the dark web he would not have been aware of the robots.

"Dr. Harris, have your labs been successful in finding a way to prevent the onset of the fuzzy-copy syndrome or reverse it?" Sam asked.

"Not completely in either case. We have been experiencing constraints while trying to access the ship's computers during our research."

"Chairman Brock, can you help in this matter?" Sam asked.

"I think so, Sam," Brock replied.

Sam had expected the Chairman to talk directly to Dr. Harris. *This was a beginning*, he thought.

Dr. Harris smiled. "Thank you, Chairman. My people will be in touch with the ship's computer techs."

"Can you explain why we were not given a heads-up about the tranquilizers?" asked Sam.

"Yes, the original maneuver planned for the ship was not going to require a period of weightlessness. By the time we found out about the change, we found ourselves rushing to get our section of the ship ready. It was only decided at the last minute that it would be beneficial for the whole ship, not just my people. Sorry, I realize now how suspicious it must have looked. Everybody on the ship took the same tranquilizers."

Dr. Harris saw the look on Brock's face. "Chairman, you can evaluate a cross-section of the crew until you are satisfied with the results."

Now it was Chairman Brock's turn.

"The food... what have you and Anastasia done to the food you have been supplying to the ship?" He was almost shouting when he finished his accusation.

Dr. Harris was adamant. "Nothing. Neither I nor my people had done anything to affect the food."

Brock was going crazy. "What? Do you think..."

Sam interrupted him to address them both.

"We are all on this ship together and neither of you should forget that fact. We have recently passed the halfway point. The protocol now calls for all Department heads to begin planning for the end of the voyage. I will return to an active role as Chief of Operations as per the voyage's plan."

Sam made up that part. No one questioned him about the plan because they could not remember one. Actually, the Captain and Dr. Harris seemed relieved. They both would take this halt in hostilities to plan for the other's demise.

CHAPTER 17

Earth's surprise

Thomas Hacker and his computer techs, who were versed in malicious software, entered. Everyone was surprised when they realized that a group of Dr. Harris's technicians were mixed in with them. Sam and everyone in the room were speechless.

Tom began to explain what was developing. Before he even finished a sentence, Sam interrupted him.

"We need to have everyone from both sides present to hear this together so there will not be any questions about this latest information."

The Captain turned on the ship-wide intercom to allow everyone to hear firsthand. Tom started over again:

"About 225 shuttles were entered and reprogrammed when the nuclear bombs were installed. These shuttles were launched and eventually managed to destroy most of Earth's pursuit shuttles. But we think twenty did manage to survive. To prepare for the flip, a large number of shuttles had to be relocated to the vacant shuttle parking docks. The empty parking docks were created when the shuttle attack force was launched against the Earth's pursuit ships. To move the shuttles around the hull, a two-man crew must enter the shuttles. This allows the air from the shuttle and the ship to mix. The recent shuttle entrees were the first in almost 350 years."

Everyone was listening intently.

"We have identified the food pathogen!"

Everyone started talking at once. Sam did not have a gavel, so he borrowed Colonel Hooker's pistol. He banged the butt of the pistol on the desk, but it still took about 10 minutes to regain order.

Tom raised his hands.

"You must all shut the hell up," he said. "Captain, do you remember the last radio transmission from Earth's fleet?"

The Captain had a puzzled look on his face.

"According to a computer analysis of the word cadence of the pursuit fleet's last radio transmission, they cut it short. The voice message had an encrypted software program embedded. We only recently cracked the code. I will get back to the code in just a minute.

"Our initial analysis indicates the pathogen is a form of Bacillus Anthracis."

The room broke into a pandemonium, and people were losing it. Finally, Tom had everyone sit down and listen to the rest of the story.

"The Anthrax spores appeared to have been genetically modified," he explained. "We will soon have an answer. The labs will do further testing. Now, back to the shuttles.

"When the first sign of infected food appeared, swabs were used to collect samples from all over the ship, including the open shuttles. We were trying to identify where the greatest concentration of the anthrax might exist or where it might have originated. Well, it exists in the shuttles. A thorough examination of the shuttle's computer software revealed the presence of the encrypted code in its hard drive. The code is embedded only in the shuttles and has not been found anywhere else on the ship. This is the reason we were never able to locate it.

"This discovery caused us to perform a detailed physical examination of the shuttles. We found hidden equipment in place to store the anthrax, and it was released when the proper code was received. The last message from Earth was that signal. The fact that no one opened the shuttle early on saved everyone on the ship.

"So, why are we all not dead? Fortunately, the antidote was also included and was released into each shuttle. We think the antidote was released around 100 years into the voyage; it was a default time setting. This has allowed the anthrax and the cure to react with one another in an exceedingly small space for about 250 years. The cure accomplished its job just as it would do in a person's body. The anthrax has lost almost all its potency and will not, we hope, be a threat to human life except maybe for the weakest among us; if it is detected in time, a person can be cloned and not suffer any effects from the pathogen."

Both Tom and Dr. Harris's chief research scientist were at a loss to explain why Earth had put a cure into play. What possible reasons could there be? Everyone awake would be dead. The question seemed moot. There had to have been a reason, but it looked doubtful if they would ever find out.

Tom closed with one last fact concerning the encrypted computer code.

"We decoded the sections of the code we had managed to find and reveal. We found fragments that we had not been able to decode at all, and so some of the mystery remained. Perhaps one day, we will know it all."

There was a long silence. Everyone on the ship realized exactly how close they had all come to dying. It was only by pure chance that no one had opened a shuttle after the anthrax release code was sent and a shuttle was not opened before the cure had managed to weaken the anthrax.

Dr. Harris spoke up. "A large part of the crew had not eaten in quite a spell. She surmised that when everyone resumed eating meals. There are going to be several people who are extremely sick. The ship does not have a large medical facility or a large supply of medicine on hand. I suggest that as soon as anyone becomes seriously ill, we immediately put them in cloning storage. We should halt all unnecessary cloning to make all cloning equipment available for this purpose. I also think my people and the Colonel's people work together to develop a DNA change that would make us immune to anthrax."

The Colonel reluctantly agreed.

Sam realized the two groups were two blocked by the anthrax, and neither had a clear advantage now. The Colonel had complete control of the ship. With his security men spread out throughout the whole ship he clearly had the advantage now over Dr. Harris. The Colonel was beaming now with a large amount of smugness.

But Dr. Harris was not quite finished.

"Colonel, your men are welcome to stay on station to help maintain security in my section of the ship. However, I expect to see them disappear as soon as things return to normal. Are we agreed?"

The Colonel never had time to answer.

"Captain, is it not time to turn off the ship's intercom?" asked Sam.

"Yes," replied the Captain, and instantly turned off the intercom.

Was broadcasting the proceedings a smart move? wondered Sam.

CHAPTER 18

Sam Makes His Move

Sam believed he had finally gained the upper hand. He wanted to have a private conversation with Tom. They decided on Sam's quarters, which were surveillance-proof.

Sam realized that he was taking the biggest chance in his life of almost 400 years, ever. Tom arrived on time; they settled down.

"Tom, would you like to share a bottle of 500-year bourbon?" Sam asked. "We have to make some serious decisions."

"We sure do," said Tom, nodding solemnly.

They sat there for a long while, looking at one another and enjoying the bourbon. Finally, Tom broached the subject concerning the present conditions aboard the ship.

"How do you feel about helping me bring about some changes aboard the ship? I am talking about picking winners and losers."

"It is funny you brought the subject up," Sam replied. "But before we do, how about sharing with me some details of how you got involved in the Project."

"Quid pro quo?" Tom asked with a smile.

"Certainly. Tom, tell me something about your early life and when did you first become aware of the Project?"

"Normal, two parents upbringing on a cattle ranch in Wyoming. I graduated from MIT with PhD in generative AI and using computers to control cloning. I was hired as a mid-level computer scientist for the original unmanned probe project and, as you know, it was scrapped for a number of reasons. Some of the scientists were rehired for the migration project. The competition for these positions at the top level involving hardware and computer programs was brutal. I decided to undergo black-market cloning. This would allow me to start over as a new person with a new name. We purposely changed my DNA. Being the computer sleuth that I am, it was relatively easy to access my personnel records and make the necessary changes. Then, it was easy to get noticed. I was over-educated and easily got a job as a simple technician. I was hired to help with the computer design and installation on *New Hope* and the shuttles. This turned out to be the best possible path to allow me to make the final cut and become a member of the crew."

Tom took a deep breath and another sip of the bourbon.

"But what about you, Sam? You were young when you had your first cloning, and it was years later before I met you on this ship. The cloning techs had never met or heard of a person with severe hyperthymic syndrome (a rare type of savant). When we were asked to remove all the visible signs using cloning, no one thought anything about it. At that time, everyone regarded it as an opportunity to gain clinical knowledge.

"But I wondered why? I pulled my work records involving my participation in your cloning. I was in charge of the temporal lobe section of your brain. It mostly deals with a person's short-term memory; I made a few clicks on the computer, and I disappeared from your memory. The only drawback in doing this is the fact that the subject does not remember anything during the period when you erase specific memories. When you had the Captain send for me to report to the bridge, I thought maybe you had remembered. It was not until later that I knew it was the Captain all the way. I have watched you for over 300 years. You have never in your normal speech had to stop and search for words, nor said 'huh,' 'aah,' or any sound that is used to stall in a person's normal speech patterns. Only recently have I come to understand what a big advantage you have when you are dealing with people who have a normal memory."

"Can you tell me about Dr. Harris and anything she may have accomplished that might stop us from taking her down?"

"Dr. Harris has made progress that far exceeds what one would expect to be possible in the field of cloning. She and her people are working continuously to clone and create a class of people who will be happy to maintain their station in life and perform the work necessary to run the day-to-day operations of the ship, especially the cultivation of crops and raising farm animals. They will all have a cloned mate. This all reminds me that Colonel Hooker is doing the same thing with his security force.

"I have more freedom of movement than anyone else aboard this ship. The danger of a pitched battle grows more remote every 24-hour cycle. The big question: I have observed changes that boggle the human mind over time. In the beginning, there was a lengthy period of cooperation that involved everybody on the ship. The manner in which the ship escaped produced a team attitude and a sense of belonging even after we were responsible for the deaths of thousands of fellow humans. What happened then even surprised me.

"When it dawned on people that there were not going to be any repercussions for the ship's actions, they began a march toward small-mindedness, and the attitude me, me, only I matter. This is happening now. There are many dark and lonely areas on this ship where unspeakable things are happening to people, especially women and children. Yes, there are a fairly substantial number of children hidden away on the ship. Brothels are rampant. The ship has a vibrant black market; if it is on this ship, you can get it. The ship is evolving into a middle-aged Earth society. There is a large group of men and women who no longer behave as if they are members of the crew. This is kept out of sight as much as possible.

"The introduction of drugs has played a prominent role in the decline in the conditions on the ship. It was not until recently that I understood why the different sections were taking such a hard line on people's comings and goings. They were and still are trying to prevent drug smuggling. What is happening on this ship now mimics people's behavior all throughout human history."

Sam interrupted him. "What you are telling me is that we now have three distinct factions on this ship. Does anybody know who the leader of the drug cartels might be?"

"No one knows for sure who they are. There are rumors about where the main operation might be located. But there have never been any activities observed on either one of the camera networks large enough to fit the bill. So, we really do not know much for sure."

"Tom, how close do you think the Colonel is to locating the drug cartel's operational center?"

"No, not close at all. Can we put the drug question on hold? Quid pro quo, remember?"

Sam smiled. "Alright. Born in Washington, DC. Two parents present; mother, PhD theoretical physicist; father, career government bureaucrat. I spent my early years trying to appear normal. When that did not work, and people around me were trying to stop me from advancing in my career, they would just disappear one day. I never got caught or even accused. I became particularly good at making people disappear, and I still am. You have firsthand knowledge of this, I believe," Sam said, looking at Tom directly in the eyes as he spoke.

Tom gave absolutely no hint that the message he just received was a clear threat.

Sam continued. "I was the last director of the American bureau that managed black opts. We actually did not have a name. I still do not know who actually knew of our existence. During this period, I got the nickname 'Chief S.O.B.' because of my actions as the head of black opts. Then, I was offered the position as the assistant to the then Chief of Operations, Samuel Otter Bernard."

Tom had a look of utter horror on his face after what he had just heard.

"I had been hired to assist the chief and be ready to assume his duties if the cloning was a failure. I had been in the job for almost one year before the Chief's cloning. This had given me enough time to remember everything the Chief had experienced in his life, especially names and dates. The Chief had wanted to come back as a much younger person. This eliminated the lookalike problem. Even in the early days, changing your DNA was easy to accomplish. The Chief did not survive the cloning. He had died just before the cloning started. His essence was lost. There were only two techs on duty when he died. They had been well paid by me and wheeled me into the cloning laboratory, and I took his place on the gurney. I now had his DNA and all his memories down to the smallest details committed to my memory."

Tom looked dazed.

"The cloning records show everything went perfectly for the vanished Chief. My cloning records for that day do not exist. Within a year or so, the three techs all died in a convenient shuttle accident. Need I go on, Tom?"

"No. That will not be necessary."

CHAPTER 19

Conspirators

"Do you want to work together as a sort of alliance?" asked Sam. "We will have to rely on each other for our safety and develop some way to communicate secretly."

Sam released a trial balloon. "Have you ever heard the rumors concerning the existence of a dark web on this ship?"

"Yes, but I have never seen any evidence that the rumors are true," replied Tom.

Sam looked at Tom, trying to get an idea if his last statement was true.

"I have searched on a regular basis for almost 300 years and never found even a clue, much less that it exists," Tom replied.

Sam knew that by playing his strongest card and the desired results were not obtained, it would mean the end for Tom. Tom also realized this.

"Tom, there is a dark web, and it is more than simply a way to communicate. We must keep this between us and only us.

"We are going to, in the coming years, have to enlist a lot of help. The first thing we need to do is keep Brock, the Captain, the Colonel, and Harris from trusting each other. Drug people are another matter.

I do not know how they will fit in the mix. We should put the three of them on hold and take no action," said Sam.

"How do we convince the Colonel to keep his men on station throughout the entire ship?" asked Tom.

"He may not realize it yet; he has at this moment what he has been seeking complete control of the ship. Controlling him should be our first objective."

"Who do you suggest we try to enlist?" asked Tom. "You realize that anyone we approach, and they do not agree to enlist in our takeover, we will have no choice but to murder them. The candidates must be smart enough to realize that refusing our offer will be a death sentence. They will only be joining us because they do not have a choice. Their loyalty will always be in question."

"I do not mean making them disappear by just putting them in cloning storage. I am talking about destroying their mind and body," said Sam. "The first thing we are going to need are secret enforcers, who will follow our orders and ask no questions."

"We need to establish a third cloning lab in secret," proposed Tom. "With a resource like this, we could custom design each enforcer to suit our needs. Then, we can design mates for them with strong maternal instincts and have each female pregnant when they come out of cloning. This would give us iron-clad control over the couple by using their young children's continued well-being as a threat. I will get to work on this ASAP."

Creating a group to enforce their rules and laws in secrecy reminded Sam of the Danites created around 1838 in the USA. Many people from those times thought of them as a very cruel lot, but not the people for whom they were working. This would be as true today aboard this ship as it was then.

"I want to try to enlist Sergeant Sawyer," said Sam. "Then his wife and two children would be awakened if he agreed to join the conspiracy. They could control him completely just as they would the enforcers that would follow."

The next move Sam brought up was to create a court system to manage ordinary criminal situations. He would be the chief of the court and the only judge. Assuming he joined them, Sergeant Sawyer would be the head law enforcement officer and be promoted to the rank of Major. Based on past conversations with Dr. Harris and the other group, they both believed they could get them to agree. To accomplish this, each faction was going to have to be convinced they were getting something extra out of the deal, and the other side was not.

"Well, there is no better time than now to ask Sergeant Sawyer the question," said Sam. "I'll call him."

He soon arrived and took a seat at the table.

Sawyer had been cloned several times. In fact, I'm not sure if this man is even Toby Sawyer, Sam thought. *But we'll soon know the truth if we wake his wife up.*

Sam and Tom explained the job they had in mind for him, and he would be promoted to the rank of Major.

"Am I going to be in charge of the judicial system's police officers – the enforcers?"

"Yes. Completely under your control."

"When does the new rank become effective?"

"Immediately," replied Sam. "So, are you going to join us?"

"Yes."

The next words out of his mouth answered Sam's identity question.

"Will it be possible to wake up my wife and two children?"

"Yes, Major. "We can almost immediately, but to save you from unnecessary cloning, you need to coordinate with Tom here to bring your family members back to the proper chronological age to match yours – unless you want to change your age."

"I am going to need time to get a third cloning lab up and running," Tom interjected, as the two of them had agreed beforehand.

CHAPTER 20

The Setup

Sam himself went to see Brock, the Captain, and the Colonel. He went over his plan to form a police force and a court system. He would be the chief judge.

The only reaction he received from them was cold stares. He had expected this.

Chairman Brock put their feeling into words. "What is in it for us?"

"A proper legal system will make it much easier for the three of you to control the ship," said Sam. "Colonel Hooker, I heard you have been getting a lot of grief from the ship's company about your men on station throughout the ship. They have all essentially moved to permanent quarters at their duty stations. This has placed a strain on other people's resources.

"Chairman Brock, in your case, the ship's operational manual spells out everyone's duty. I wrote the operational orders, and they do not mention any duties for you or the council until we disembark at the end of the voyage. But I see now I was wrong. Someone with your administrative abilities needs to take charge of all things that do not directly involve the ship's day-to-day operation.

"Captain Volkov, we have had conversations in the past about the ship's operation. Please stop me if you think I am about to disclose trust. The bridge crew is probably the smallest and tightest-knit group on the ship. This court system will not infringe on your authority as the Captain; you will continue to be the final decision-maker. You will be able to overturn court judgments and commute or pardon all sentences as you see fit. You, as Captain, will still have complete access to every area of the ship and the computers. I believe it will affect you the least."

The Colonel finally asked the big question.

"Where are Dr. Harris and her group? Why aren't they here today?"

"I am not going to ask her permission," said Sam. "I am going to inform her that what we have agreed on is final, and there will be no changes."

Sam hoped he had figured this out correctly. The Colonel looked at him the way the lion looked at a wildebeest. He began slowly, choosing his words carefully.

"Do you plan to go to this meeting alone?"

"Yes, I do."

"Are you not approaching this with a somewhat cavalier attitude?"

"I do not think so; your security people are in her area, are they not?"

"Dr. Harris has not given them full access yet. You will probably be by yourself. Let me send someone of a higher rank to accompany you. Major Sawyer seems to get along well enough with you. Do you know any reason he cannot accompany you?" the Colonel asked.

"No, none," replied Sam. "I trust him completely. I will ask him if we should bring extra security."

So far, it is going better than I had hoped, thought Sam.

"We three need to have a private conversation about your proposal for change. I will get back to you ASAP. We realize time is not on our side," said the Colonel.

***

Sam met with Tom and Major Sawyer to discuss the meeting with them.

"I believe the Captain, Colonel Hooker, and Chairman Brock swallowed my story completely," revealed Sam. "They will have an answer for us soon."

Next, Sam and Tom explained the existence of the dark web to the Major. Nothing seemed to surprise the Major. Sam got the web running. Sawyer was watching and almost made a comment when he saw Sam enter the first 99-digit code from memory. He did make a comment about the 100-digit second code when it was entered.

Major Sawyer looked at Sam. "Have you remembered everything about me from our first meeting almost 400 years ago?"

"Yes, I have everything," Sam replied.

They began to watch the dark web's surveillance cameras. Dr. Harris was discussing a first-hand report about Sam's meeting with her staff. The identity of the spy had not been revealed, only a code name. The name Tiny did not ring a bell.

This new development had really thrown them all for a loop. They were all drawing a blank about what their next step should be.

Finally, the three of them all agreed on one thing they needed to delay the meeting with Harris, for as long as possible or at least until the court system, police, and a brig were all in place.

They were interrupted by a video call from the Captain.

"Everyone, get out of the room before I take this call," instructed Sam.

The Captain appeared to be alone in his ready room as did Sam.

"Sam, I have serious reservations about approving your plan. These are substantial changes you are planning," said the Captain. "Explain to me what you plan to do first."

"I plan to go after the drug cartel first. They are a problem," Sam replied. "I think everyone will agree, they need to just disappear. I plan to ask the three of you for your permission to clone officers for the police force. I estimate I will need 50, but I don't have a final number yet. I am going to need to get Major Sawyer's input first. Maybe you could loan me Tom to head up the officers' cloning? Does he still manage the computers used in your security department's cloning lab?

"We can set it up where both Tom and the Major report to you. You can then decide how involved you want to be. I will make that point truly clear to both of them. I would feel more comfortable working with the two of them. You know them well, and I think you will get a better picture of what is going on in their respective areas of operation if they and I keep you up to date."

"Okay, Sam, go ahead. I will inform Brock and the Colonel of my decision," said the Captain. "I assure you they would not be a problem."

The Captain closed the video link, and Tom and Sergeant Sawyer returned to the room.

CHAPTER 21

Mutiny Planning

They were extremely happy with this turn of events. Convincing the Captain would be the hardest.

The first thing they needed to accomplish was to wake up and clone Major Sawyer's wife, Janice, and two children, David and Alice.

"No one knew more about the ship's computer system than Sawyer's daughter," Sam told Tom. He took a minute and explained her role to Tom leading up to the launch.

"We are going to need a place to call headquarters," said Tom. "It would need to be secure from electronic eavesdropping and the physical location a secret."

The three of them began a round table discussion about where they could locate the entire infrastructure. Sam was the first one to produce what might be the answer to the question at hand.

"Look, between the three of us, we know every cubic inch of this ship," Sam said. "There is one large area no one ever thinks about, the space between the inner hull and the outer hull."

The Major and Tom's eyes widened. They did not know what to expect next. Sam began a rather long explanation.

"Meteor penetration was a real possibility for a journey lasting this long. The engineers spent countless years trying to come up with a solution. An inner hull was that solution. When designing both hulls' thicknesses, it became apparent the safety requirements were impossible to meet. The solution to the problem turned out to be rather simple. Just add two more for a total of four hulls. That would mean three hull voids.

"The total distance between the innermost hull and the outer hull was designed to be approximately three hundred feet. The three voids would be sealed off into smaller sections. Hatches would permit full access to every area of the hull voids. A last minute add on was to fill the entire inner hull voids with plain old H_2O. Water happens to be one of the most efficient ways to protect living tissue from radiation, and three hundred feet of water would protect the crew and equipment from normal cosmic rays and even from the most powerful solar flares humans have ever observed. There are hundreds of thousands of cubic acres feet of water stored. This should buy the colonists time if water is not really available upon arrival. A meteor large enough to penetrate into the ship's environment would destroy the ship with or without the beefed-up hull. This unused space will be the perfect place to set up the court system."

"I believe the best place to access the outer hull and construct the facilities needed would be in the wide-open spaces located in the middle area of the ship," said the Major. "I am familiar with this area of the ship."

They were going to need a security detail. Sawyer believed he could round up a couple dozen men who were trustworthy.

"I will be out of reach for the near future," he said. "I plan to explore the middle decks extensively. I will send progress reports back to both of you as necessary. The squad and I will be operating in Anastasia's backyard. If we are discovered, how do you suggest I manage the situation?"

"You only have two choices; the best would be to try and explain it away," said Sam. "Do you remember the shuttle technician that went missing around the time of the flip? Well, she disappeared in Anastasia's area. The evidence we uncovered points to her being detained and then scanned before they killed her. As far as I know, they are not aware we know the truth about her disappearance. It should not be that big of a problem convincing them. Just say you are trying to put her disappearance to bed.

"Then, there is the second choice. However, it is not to be used unless you can contain the peripheral damage created when you kill them. We can get Tom to render the bodies and add them to the ship's biomass. This will prevent anyone from being able to clone them; their minds will simply no longer exist."

After the Major left, Sam and Tom sat down and stared at one another for a long time. They knew what was going to be required involving the cloning of Sawyer's family. They were about to commit the first betrayal of a comrade. And there were probably going to be more betrayals.

"I have been involved in the last few cloning sessions of the Major," said Tom. "His conception of what actions are right and which actions are wrong are about the same as everyone else's. We are going to alter his family in order for them to survive on the ship. We will be doing it in their best interest. There is a likely chance he will realize what we have done. Even if he does, making a problem out of it with us would be the end for him, and he knows this."

"Tom, I am going to need Alice as soon as possible," said Sam. "We need the Dark Web for assorted reasons. With her knowledge of the ship's computers and surveillance systems, we will be unstoppable."

"I will see to it that she will be properly prepared for her role after the cloning," replied Tom.

"Tom, I am going to be busy for a long time," said Sam. "I plan to spend time with the Captain. He needs to feel like he is in control; keep me up to date."

Sam contacted the Major. "Let me know when you are ready to explore the inner hull.

***

Major Sawyer had been a busy man. He had put together a small force of ten men and two women.

"I can be ready as soon as right now, but we may be getting ahead of ourselves," he replied. "A visit to both Anastasia and Qiao Hui is needed before we go exploring. It will be much easier with their cooperation if we can get it."

"They share the same area, and from what my men on the station have observed, they are close," reported Colonel Hooker. "They even share the same living quarters and have for hundreds of years.

"Qiao is the only one with a working knowledge concerning the inner hulls. The primary L.O.A (loss of atmosphere) alarms are located in the maintenance area. There had been meteor hull impacts since launch. None managed to penetrate the first inner hull. None have been larger than grains of sand.

"Qiao has a whole crew trained to repair meteor damage. We will have to get their help to open up any sections. There are ongoing repairs now. We will not know how extensive the repairs are until we see them for ourselves.

"Anastasia, on the other hand, is going to be a complete mystery. No one outside of her department has talked face-to-face with her since the ship launched, of which I am aware. So, I am going to put together a larger force and hold them in reserve. This could delay you."

"Major, you and I need to make a cursory visit to their area with only a small escort, maybe two men," said Sam. "We need firsthand intelligence. So please contact me when you are ready to go."

Sam wanted to research both their records. What better way than in his quarters? He would not need to see any files.

Anastasia had been recruited very aggressively. She was the top botanist in the world and also had a PhD in plant DNA. She had been successful in growing plants in some of the world's worst soil and climates. For generations, her family had been successful as farmers. There was only one problem: their primary crops had been opium poppies, marijuana, and Erythroxylon coca plants. These facts had been well hidden.

Sam knew only about this because he had been the last director of intelligence for the USA. He was, in effect, the last one in the world. The United States of America had for all intents and purposes taken control of the entire world. When Project Migration started in the name of saving humanity from itself, all the world's resources were directed to it. This caused shortages all over the world, and billions of people died.

What have I done? thought Sam in a moment of weakness. Then, in a flash, he knew. *I saved my ass.*

Sam had no such memories about Qiao; in fact, he had none at all of him before launch. This would be a starter project for Alice.

***

Time seemed to be flying by. As somebody once said, we are over the hill; now it is downhill from here.

You could tell things had changed. When the voyage first started, time was measured in how long it had been since launch. Now, time was measured according to how many days were left before we got there. The Gregorian calendar had been forgotten. People were beginning to ask questions about the future after they got off the ship. The bridge had been silent on the aspect of the migration approaching its end.

The Major finally contacted Sam about the fact-finding mission. When the day arrived, they met in the Major's new office adjunct to the bridge. The Captain had insisted, and they both figured Volkov wanted to keep a close eye on Major Sawyer, who was now the Chief Law Officer aboard *New Hope*. The Captain had taken an interest and joined the Major and Sam.

"Captain, we want you to contact Anastasia and Qiao and give them a heads-up about the upcoming visit," said Sam. "However, give them as little information as possible."

Sawyer and Sam had discussed the Captain's involvement earlier and how to handle it. Mostly, they wanted to keep him in the dark about their plans.

"How much advance warning should I give them?" the Captain asked.

"We do not want to be the ones to set the date. They are going to need time to get their house in order," replied Sam. "This will be the first... let us just say inspection, in the last 350 or so years. Just tell them they should notify you when they're ready to receive us."

Sam could tell from his reaction that the Captain was not too happy. Maybe he did not understand what was behind his reasoning.

"You see, Captain, we do not want to have a confrontation without knowing how it would end. The four of us will be greatly outnumbered, so we are going to be especially nice until we find out more about them, meaning how dangerous they are. We will not spare a single soul if that is what is necessary for us to take control of the ship."

***

Eventually, the day arrived. Sam went to the bridge and met Major Sawyer and two other men. One was a structural engineer chosen by Oliver Whipple, and the other a computer whizz chosen by Tom.

They were dressed to look like ordinary security personnel. This had been Sam's idea. He hoped that by giving trained people a look at the objective, a workable plan would be easier to formulate and execute. They decided to go armed only with side arms. Sam would be unarmed.

Colonel Hooker produced an emergency plan. Hundreds of entry points allowed access to the area from the secret security corridors. Their hatchways were hidden from view. Hooker and the Captain proposed to station groups of armed security personnel at each entry point. This would give the four envoys almost instant protection in case of trouble. The mind trap was used to protect each of them in case they were killed.

Everything was in place and so they went using the so-called public corridors.

Sam was taken aback when they arrived. Anastasia and Qiao both appeared to be the perfect picture of health. They had to have both been recently cloned. They had plenty of time to get ready for the visit, including cloning themselves.

Sam immediately realized something was very wrong with what he was seeing. When the ship launched, there were only two departments on board the ship that had cloning facilities. Anastasia and Qiao's department was not one of them.

They all exchanged friendly greetings, and everything seemed to be going well. Major Sawyer was glad he had waited for a plan and had not just barged in on his own. Every other person was armed and looked like they were ready to deal with anything that might come up.

"Is there a place where they could sit down and talk?" asked Sam.

Anastasia showed them a table next to the ship's outer hull. Sam tried his best to exhibit a calmness he did not feel.

"Have you heard about the changes in the ship's disciplinary force?" he asked. "Law and order were on the way."

"Yes, we have heard everything about your plans to become a one-man judge and jury," replied Anastasia. "I also think I know why you are here. You are looking for a place to put the police station, courtrooms, and prison. We have no unused space, which would be large enough and available for your use. You are wasting your time here. I can have you escorted back to the border when you are ready."

"Before we go, there is one thing we need your help for. During the ship's flip preparation, a computer tech was lost in your area," said Sam. "We are hoping you might be able to shed light on her disappearance. Hopefully, you can show us where she was lost."

Anastasia shook her head. "None of my people were there during the time of your tech's disappearance. I have nothing to add."

"Do you mind if we go to the area and have a look?" asked Sam. "We will need a guide to show us the way."

Qiao spoke up. "We can do that. Just relax, and I will have one of my men take you to the exit port connected to the shuttle in question."

While they waited, they were served beverages. *They had never acknowledged that they knew where she disappeared,* Sam thought. *Now, they were going to take us to the spot. They are lying, for sure. This is a trap.*

Should they alert the security people waiting in the corridors for the worst? Now, this might not be the best time to act. With an exchange of glances, Sam and the Major agreed to wait and rather follow the guide.

After a few minutes, the guide arrived and took them to the correct shuttle. They made a show of examining the site. They returned to the headquarters area and soon left and returned to the bridge. No goodbyes were exchanged.

After debriefing everyone it was decided to destroy Anastasia and Qiao as soon as possible. The facts they had uncovered spoke loudly about how out of control the two were.

It was apparent they had access to the codes the dead tech had in her memory. The most important discovery was that they had a cloning laboratory. What they were doing with the lab was anybody's guess. The Captain's gang met and made plans.

CHAPTER 22

Speeding Up the Takeover

"I will have the Sawyer family up and running soon," Tom announced. "When Alice is ready, she will be turned loose on the computer. Also, I think I may have a way to explore the newly-found cloning lab before the attack. I will get back to you."

"All I need is a little time," said Colonel Hooker. "Then I will be ready to take over Anastasia and Qiao's operations. I want to capture both of them alive. This will allow me to gather all the information and memories contained in their conscious and unconscious mind before killing them."

Oliver Whipple was put in charge of building the infrastructure for the court system. He would begin making plans to access the inner hull voids, locate the necessary construction material, and start work as soon as he got clearance.

It was decided to attack Anastasia and Qiao as soon as possible. Colonel Hooker took the floor.

"My top officers and I have entertained several approaches. The first and most dangerous is a plain old frontal assault, entering the section using the security tunnels. The attackers will have complete surprise on their side. The marines involved want very much to engage in combat. They had all been genetically engineered to be vicious killers but had not been allowed to practice their craft yet. We will not be able to ignore their will very easily.

"The second approach is much simpler. The area occupied by Anastasia and Qiao can be sealed off, and the atmosphere can be vented into space. This would kill everybody. There will be no loss of life among the friendly attacking troops."

Everyone agreed the frontal assault was the way to go. There were too many unknowns about damage to the flora and fauna if you vented the section to open space. There would not be any prisoners to interrogate. That fact alone sealed the deal.

D-Day finally arrived. The Colonel had managed to put together an attack force of 400 men and women. They were all in place and awaiting orders.

Sam and the Captain were somewhat surprised by the relatively small number of troops.

The Colonel smiled as if he had been waiting for this very reaction. "Oh, by the way," he said, "there will be 150 robots from the ship's engine room. They have all been modified for combat with humans. No human or any animal we know of can defeat them in physical combat. They will all be armed with the same weapons as the live troops. Our live troops will all have a transponder and a visible symbol to identify them as friendly. The robots can survive a larger amount of battle damage than humans before they are defeated.

"Last but most importantly, we have kept this recent development concealed. The medics will have to make a quick decision when called to a wounded soldier. They can treat all nonlife-threatening wounds just as medics have always done. But when they encounter a life-threatening wound, they are to call in a cloning tech. We have developed a portable cloning mind trap. The wounded can be saved on the battlefield; there will not be any need to take their bodies back to our cloning labs in security. This development has given our killers a sense that they are now invincible. This treatment will not be extended to any combatants on the other side except Anastasia and Qiao."

The Captain, Sam, Whipple, Tom, Brock, and Major Sawyer had all gathered on the bridge to watch the attack unfold.

The Colonel had decided to participate and entered through the nearest hatch, accompanied by twenty-five men. Something was wrong; there was not a soul in sight. His training mandated that he does not allow all his men to gather in one spot.

"Everyone, hold your positions," he ordered. "I need Intel. All squads send out scouting patrols. Be very careful and keep your cameras and mikes on."

Time seemed to be dragging on. There was total silence. Then, one patrol mentioned what looked like a well-worn path that went up to an outer wall of the ship and stopped.

The Colonel then decided to have his men advance out into, for lack of a better word, open countryside. This had the effect of exposing his men to the enemy. Sixty percent of the land was heavily wooded. Anastasia and Qiao had prepared fortified positions, giving them various fields of fire and turning the open areas into kill zones. There was not sufficient cover for his men, and slowly, they were systematically being killed.

"Join the battle!" the Colonel ordered the troops who had been left at the entrance points.

When they did, the hatch at the end of the well-worn path opened up, and dozens of armed men poured out. When they joined the battle, too, the Colonel realized they were now in a crossfire. It was time to call out the reinforcements, and he did.

In just a matter of minutes, the robot fighters joined in. The enemy was powerless to stop them. They killed with wild abandonment, and within 45 minutes, it was over. Both Anastasia and Qiao were found wounded but alive. The mind trap was employed, and then they were allowed to die.

The Colonel had realized that they had not encountered any non-combatants. He sent a patrol to explore the area behind the hatch where the enemy's second wave exited from.

In just a few minutes, they radioed a short message, "Colonel, you need to see this."

They had found the home of the cartel. There was a large drug lab with quite palatial living quarters located in the inner hull.

The workers and the spouses of the men killed in battle were all there. The Colonel gave secret orders to separate the workers and their families from everybody else.

"Over the next few days, take everybody else. Use the portable mind trap, and then move them to cloning storage," he ordered.

The workers were eager to help, hoping to save their lives and their families. They willingly showed where the drug crops were planted. There had always been different areas on the middle decks that the surveillance cameras never seemed to show completely. Now the group on the bridge knows why.

The Colonel spoke directly to the ship's First Mate, Oliver Whipple. "Have you got your people trained and ready to take over food production and maintenance? You need to get the transition accomplished as soon as possible."

The colonel was surprised at how quickly the workers were willing to cooperate and share all the information they possessed about the drug operation in return for a promise to be awakened if they were put in clone storage. He joined the group on the bridge.

They all hoped that today's actions would prove to be positive, but several developments worried the Bridge Committee.

The number of people found to be living in Food Production and Maintenance was more than double the number expected. A sizable percentage of the cartel workers had children of various ages, seemingly all coming from a normal birth without any of the cloning traits in use at the present time.

This raises the question, what else is taking place of which we are not aware?

Anastasia would not have much of a problem hiding extra food production for her people. Most of her people would have been terribly loyal. They depended on the cartel for their drugs and everything else.

"The cloning laboratory has five cloning chambers," the Colonel reported. "The serial numbers indicate they were among a group of twenty-five that were removed several days before launch and replaced with the newest model. Apparently, these five never made it off the ship."

"For that to happen, there had to be a conspiracy already in place," Sam muttered to himself.

He slowly looked around at blank faces. *It might still be in place*, he thought.

"Captain, we have one particularly important question before us; what do we do about the drug distribution network?" Sam asked. "We do not know the number of people who are addicted. We cannot take a chance and have the worst-case scenario occur, and that would be to have two thousand people go cold turkey all at once. If everybody agrees, let us provide Oliver with all the help and resources needed to keep the drug business in operation.

"Don't you think it is odd that there was not any commercial activity or currency on board this ship? Why is there even a drug cartel in the first place on the ship? A modern recreational drug was not addictive, but Anastasia and Qiao had reintroduced the addictive drugs to the addicts. This explained why the cartel had been able to control so many people for so long. Cloning would normally rid the person of their addiction, but in this case, the drug supplier was cloning his customers. It all made perfect sense now. Clearly, they had big plans for the drug business on the new planet."

Sam sighed. He knew why they had allowed the use of drugs for years. It was an escape from reality for people. They did not really need any other reason, and as long as the people in charge allowed it to happen, it would take place.

In *New Hope's* case, the ship's officers were, in effect, drug lords, and it looked like they were going to continue to be. Even knowing the drugs were addictive and the chance of death from overdoses just went off the chart.

Knowing this did not change a thing, Chairman Brock gave orders to change nothing. "Continue to operate the same way with the same drugs," he ordered.

This all sounds too familiar, Sam thought. *It's déjà vu all over again. But it might be worth it that the drug users will certainly be easy to control in the future.*

Anastasia was a sly old girl. The cloning labs had given her an enormous advantage and made her a tough nut to crack. Not a single one of her fighters tried to surrender. They all died fighting. Sam was sure they would have lost the battle had it not been for the robots. A number of our soldiers, on the other hand, did attempt to surrender and were promptly killed by the enemy.

Sam was surprised at how quickly everyone accepted what they had done to Anastasia and Qiao. However, he realized nothing would really change until the voyage ended. The clocks never stopped ticking, and the years continue to go by.

CHAPTER 23

Changes

The bridge group, consisting of the Captain, Sam, Whipple, Brock, Colonel Hooker, Major Sawyer, and Tom, had an organizational meeting. They decided to operate as a seven-person committee, and the majority ruled. They would call themselves the Bridge Committee (BC).

Dr. Emily Harris (Medicine and Science) and Wolfgang Schmitz (Chief Engineer) had been left off the committee. The BC all felt they should gather more information before letting them join the committee or use the second option and kill them.

Tom was ready to report on the situation surrounding the Sawyer family. Janice Sawyer was cloned and had already moved into Major Sawyer's quarters. She was fine. Alice was also awake and had gotten her computer lab up and running. The lab was temporarily set up in the captain's ready room. The lab would be moved when Whipple had the new facilities ready.

"Is Qiao dead? Maybe his background history is mute now," suggested Colonel Hooker.

"We know now there was a conspiracy before we even launched. We need to find the underlying cause of what took place in the very beginning," said Sam. "The mind trap data obtained from Anastasia and Qiao did not shed any light on them being involved in any kind of clandestine operation, except the drug business. We know they had to be involved with somebody concerning the five cloning machines, but we could not find any current information at all. Based on my thinking, Alice needs to be turned loose. She is going to research all the data from the Earth files to set up a court system if no one objects. Tom, will you add that to her to-do list?"

Sam watched everyone for any sign of uneasiness. He believed the before launch coconspirators were still at large and could be sitting at this very table.

"Have you awakened David Sawyer yet?" asked Sam.

"Yes, he should be up and running soon," replied Tom.

"Have David contact me when you think he is ready."

"Tom, why is it taking so long to awaken my family?" asked the Major.

"I am just being extremely cautious with your family's wellbeing," Tom explained in his best and most believable voice.

That seemed to satisfy the Major.

Tom had some broad questions he wanted to put before the BC.

"We are in a position now to dictate to Dr. Harris how things are going to be run from now on. For starters, she will have to change her cloning protocols to suit the needs of the ship. Each section of the ship needs crew members who are cloned to perform specific duties and have unwavering loyalty to the BC."

"I do agree and will see to it," the Captain said. "I will have her contact you, and she will be ready to cooperate; I guarantee it. We are becoming a little crowded in the ship's bridge area. Thank you for expediting the Major and Alice's move to their new quarters."

"The cartel area has been full of surprises. The estimate about the substantial number of extra people present has been on my mind," said Tom. "They had to be having a negative effect on the ship's oxygen level and all the other resources. There was an extended period before there was a measurable change. The amount of flora and fauna, when balanced out, will support about 2,800 people. When the oxygen level began to drop, I assumed there was a problem with the crop, but we found everything to be normal. Over time, some of Anastasia's workers became quite friendly. They were genetically engineered to be farmers and really have no interest in who is in charge.

"After many conversations, several workers shared the reason for the oxygen level fluctuating. I was led to a remote area on one of the forested decks. Lo and behold, I was shown a plant that was built to separate water into hydrogen and oxygen. There was a pipe to carry water from the inner hull to the plant, and after processing, two pipes returned to the inner hull, one carrying oxygen and the other carrying hydrogen. As they emptied out the water, it was replaced with hydrogen and oxygen.

"They actually had done the migration a great favor. If our calculations are correct, we will have at least a 50-year supply of hydrogen fuel to power vehicles based on the number of people we have now. Anastasia and Qiao realized the substantial number of unknown people aboard would affect oxygen levels and, so they modified their LOA alarms to monitor oxygen levels. They released oxygen when necessary, and neither the Captain nor anybody else had suspected anything unusual for almost 400 years."

The BC all went back to their respective everyday activities.

***

Sam soon reached out to Tom. "We need to get together and plan a few things to help us. Can you bring Alice with you?"

"Let us wait a little while and leave Alice out of this meeting until we talk about her role," replied Tom.

"How do you want to approach her involvement in the use of the dark web? Sam, I do not want her to be involved yet.

"I agree," said Tom. "Let us get her started on producing an outline for the police, court, and prisons. I think we should give Alice and her father, Major Sawyer, as much control as they need to produce a workable system. I am still worried the Major is going to react very badly when he realizes what we have done to his family during their recent cloning.

"I am going to ask her to try and set up 24-hour surveillance on the other members of the BC. Without using the dark the dark, we may only be able to set up audio."

"I have a confession to make," said Sam. "Alice was the person that headed up the department the built the dark web. She accomplished this without a single leak and also took care of all the loose ends afterward. Meaning the people in her department, over time, all met their end at her hand."

Tom almost fainted. "You chose well," he exclaimed. "She does not even today give off any vibes that might indicate that she is capable of committing cold-blooded murder. The first thing that comes to mind is I do not think she will have any problem keeping secrets from her daddy. Let us all agree to bring her on board all the way. I will keep a close eye on her."

"Her first task should be to answer the question of what type of person makes the best police officer," said Sam. "After that, she can get busy and produce profiles on all the other duties that the ship's crew performs. When we finally get enough custom-manufactured people in place, we will then have complete control of the ship. Their loyalty will be 100 % to you and me.

"We need to have Alice find out as much as possible about Anastasia and Qiao. She has mentioned that when using the dark web and data obtained from a mind trap, it was possible to locate and identify far more points of interaction between the person in question and the world's computers."

It was taking a long time, but Oliver Whipple was closing in on finishing the new facilities in the old cartel space. He had exhibited an unknown knack for farming and food distribution. His biggest surprise was the drug business; it did not miss a beat. He left the procedures for food and drug deliveries in place. The workers were extremely happy to have him as their new boss. He was tempted to indulge himself but so far had not.

Sam was anxious to get started and asked Tom to start cloning policemen, policewomen, and spouses as soon as possible with the data available.

Conditions had become cramped in security. The Colonel agreed to reassign security officers to the new police department. This would be a start. Whipple had some of the housing finished enough to handle the transfers. He could get started now and would start using the new physic makeup shortly for all new officers. Alice was almost finished.

The Major and Janice moved off the bridge and were the first couple to move into the housing. Their new neighbor was Oliver Whipple.

Sam was waiting for his police department to become operational. He had heard nothing from Dr. David Sawyer.

Sam began thinking about how he was going to approach Dr. Harris. He could not answer his own question of whether to bring a force or try diplomacy. Convincing her to accept the new police department was not going to be easy.

He recalled the Captain's words verbatim. "I will have her contact you, and she will be ready to cooperate, I guarantee it."

Sam decided to contact the Captain and find out if any action had been taken, and then he would know how to proceed.

But first, he needed to talk to Tom. Strangely enough, he had not talked to Tom in what seemed like years. He contacted the Major and Alice, but they were not any help trying to find him. Tom, long ago, had erased his existence in the ship's computer surveillance logs.

Sam decided to arrive unannounced when he met the Captain. Alice, using the surveillance system, told him exactly where the captain was and what he was doing. Sam interrupted his lunch.

"Well, what a surprise," said the Captain. "Please sit down and join me for lunch."

"Thank you," said Sam and ordered the same thing as the Captain.

"What brings you to the bridge galley, Sam?"

"The justice system is almost in place now, and I want to plan my little visit to see Dr. Harris. My question is, can I expect her to be friendly or not? The answer will determine how many backup officers I need to accompany me."

After a rather long pause, the Captain replied. "None. You will be meeting Thomas Hacker, the new department head. Dr. Harris resigned. I took it upon myself to appoint Tom in her place. I hope you agree. I want you to make the announcement at the next BC meeting."

"Yes, of course," said Sam. "I will be glad to. But can you tell me, Captain, what caused you to act so quickly?"

"Alice and Tom have proved to be my best source of intelligence concerning the activities going on in other sections," replied the Captain. "They are proving to be irreplaceable when it comes to spying and gathering secrets about our enemies. After Dr. Harris found out what happened to Anastasia and Qiao, she was planning a move of her own. I sent a detachment of security officers during one of her sleep cycles, sedated her, and then employed the use of a mind trap. The data was sent immediately to Alice for evaluation. Based on a very quick analysis by Alice, we decided to take her unconscious body to the security cloning lab. Several dozen of her personal bodyguards were killed to accomplish the mission. The robots are proving to be invaluable and soon will change how humans kill one another. The new 'let a machine kill for you' will soon be the new human marching song."

Sam chuckled. "Captain, exactly what was Harris planning?"

"She was planning to have her Fleas rise up and take over the ship. Dr. Harris adopted the name Flea for her organization after she discovered we were looking for her organization and pretending to be Fleas ourselves. The confusion this caused made computer searches almost useless. The original plan was to have the takeover attempt upon the trip's completion and when *New Hope* was in orbit. This was an incredibly good plan.

"Harris's department controlled 90% of the cloning. They make up the rotation list of who is cloned and when over the last 450 or so years. This is the sneaky part. Everyone she put in cloning storage had her Flea syndrome installed. They kept meticulous records concerning the s she put in stasis. Instead of re-cloning the person who was put in storage, a new person is awoken in their place and cloned with her Flea syndrome. Now, every cloning produces two people with the syndrome.

"Harris had set the stage. At the end of the voyage, her computer estimated that around 80 to 90% of the active crew would be Fleas. They had developed a process using hypnosis, and it was undetectable. A coded post-hypnotic message triggers the onset of Flea syndrome. The use of the mind trap is the only way we have now of knowing for sure if the syndrome is present. I plan to keep Harris in storage until the end of the voyage and she might be your first criminal trial on the new home world."

Tom walked in and sat down with them as calmly as he could be.

"I think we have discovered the post-hypnotic suggestion," he said. "We plan to secretly imprison Dr. Harris's people a few at a time and expose them to the trigger, which we think is 'miles to go before I sleep.' It is a phase from the poem *Stopping by Woods on a Snowy Evening*, written by Robert Frost in the mid-19th century, almost 1,000 years ago. The phrase was interpreted as conveying an attraction toward death. We have not found any evidence to lead us to believe there may be more than one trigger."

"Has Alice been successful in the search into Qiao's background?" asked Sam.

"No, she reached a dead end."

That ended the meeting as far as Sam was concerned. As he got up to leave, Tom did too. They bid the Captain goodbye. They both realized bridge surveillance would not allow them to converse openly.

"I'm covered up with work, but when I have time, I will get in touch," said Tom.

"Alright. I will be spending a lot of time at the new courthouse jail complex."

***

Sam began to spend time performing his new duties. Alice had completed setting up the court system. They decided to not make public any information about the police department yet. No one knew exactly what was going to be needed to establish and maintain law and order.

The excess of people from security had been easily incorporated into the new police force. The cloning profile provided by Alice had been hugely successful. A robot unit was set up similar to a k-9 corps. An officer and a robot were paired. So far, everyone who had an encounter had been terrified. Without any armed criminals there were not going to be any problems. Nevertheless, there was a steady stream of people arrested daily for mostly petty offenses.

Violent crime was somewhat common and most always involved heavy drug users, both victims and perpetrators. The ability to clone people has had an unusual and unforeseen effect on the behavior of the crew. Violence has increased and become the preferred method to settle all disagreements, large and small. It was as if dueling was making a comeback.

Sam realized very few of the crew members even knew what a duel was. Without laws and police in place to keep order, humans always seemed to create a society where the cruelest came out ahead. Especially if no one was killed and a simple cloning fixed everyone's injuries. They were thinking about awakening a few lawyers and maybe some judges. The BC hoped this would give the court validity in the eyes of the crew. Sam hoped punishment would also restore some symbolism of a safe and civil society.

Oliver liked to share a scotch or two with Sam and discuss the drug business. The drugs they were providing on a daily basis were turning out to be having a terrible effect on the crew. When the ship's hierarchy took over the drug distribution, the need for secrecy disappeared, and any need for restraint also. Oliver had spent a long time searching for the origins of the drug cartel. Between bribes and torture, he reached some startling conclusions.

There had been stowaways on the ship from the beginning. The whole operation was started and controlled by Qiao. Anastasia was second in command, not an equal partner. Providing drugs to the ship's crew was a barter system – drugs for food. The drugs were traded in return for all the necessary services required for the cartel to exist undetected on board the ship.

"When we took over the drug trade, the barter system disappeared," Oliver said. "No one had to lift a finger to survive. It was easy to become a drug addict. We just passed the drugs out. Large sections of the ship have begun to look like garbage dumps, with people living like animals. Based on oxygen levels and how much we estimate is being produced, there are probably around 4,000 people awake, and 1,500 of them contribute nothing to maintaining the ship's wellbeing."

"This is not sustainable," said Sam. "We are headed for big trouble. We cannot produce enough resources, mainly oxygen, to sustain this number of people until we arrive at our new home."

"Alice moved into her new quarters. We made a few changes to our original plan as to how to arrange her quarters. We decided to incorporate her living quarters into her computer lab. But the most important feature is we have made it into a safe space," Oliver reported. "It can be completely cut off from the rest of the ship. Her quarters were located in the innermost void in the hull. The location was chosen on purpose. There is a shuttle attached to the hull, and the entranceway to the shuttle starts in her quarters and laboratory. As far as I know, this is the only setup like this on the whole ship. Essentially, we've turned a shuttle into an escape pod."

Sam smiled and nodded his approval, but there was one other safe space with a similar setup.

CHAPTER 24

New Faith

Sam realized that if he could see the end of the voyage, others could also. *We need a detailed meeting of all the sections to plan for the future*, he thought.

But he had one more bit of planning to do. He had to meet David Sawyer to start planning and creating the glue that would and always had held society together – religion.

David arrived at Sam's quarters. For some reason, he had a difficult and long recovery after awakening from his first cloning.

Tom and Sam had talked in length about David's experience, and they did not remember having seen this reaction before. The changes necessary to get David ready to function in the present conditions aboard the ship were enormous. The changes to his core personality made him behave in ways he did not recognize. David was having trouble socializing with anyone. He was unstable and unhappy and did not realize how he had been changed. They wondered if he was suicidal, and they realized they were going to have to confess what they had done and maybe offer him some help.

They decided to have the Major present when they approached David. They agreed not to offer any detailed information about David's cloning. The Major had to realize that major changes were made in his family's behavior. The Major needed to hear everything that came out of the initial meeting.

When they all got together, David seemed perfectly calm. For a long time, Sam and Tom used the dark web to record and achieve his every moment, awake or asleep. What they were observing now was almost unbelievable: a complete change in his outward behavior.

Since his awakening, he had exhibited the behavior of someone who was an extreme sociopath. But not today; the Major seemed to not notice any changes in his son. This was good news, but Sam still thought something was slightly wrong but could not put his finger on it. He kept his misgivings to himself since this was the first time he had been in a room with father and son.

"The big question before us was what it was going to take to get the entire crew united under one banner," Sam said. "The first and easiest solution would have been to re-clone everyone. There was not enough time left on the voyage to accomplish the task unless you operated the cloning labs wide open and dispensed with the safety protocols in place. This would obviously raise a red flag and bring pushback to an unacceptable level.:

Everyone at this point realized they had all forgotten what it was like to have any kind of time restraints. This new realization was going to take getting used to.

"Obviously, I have only been awake for a brief time," said David. "I do not have hundreds of years of experience on this ship. I do not know if I can be of any help to you in controlling the crew. But I have a plan that just might work. It was successful in the past on Earth."

Then David remained quiet. You could hear a pin drop.

"What do you want the roughly 75,000 people who have never been awakened since leaving Earth to find when you do decide to awaken them?" David asked. "There have been roughly 10,000 people rotated in and out of storage with only 3,800 awake now, another four hundred or so who are the results of natural childbirth. If we awaken the newbies too fast, they will quickly outnumber the people who have rotated in and out of cloning storage. They will take over the ship and then they might not. Can you take that risk? I do not think you should.

"Any start-up movement, religious or secular, will be hard to sell to the newbies. We can offer the existing crew incentives, power, drugs, and sex. But the biggest persuader is going to be promising them all the top jobs on the new colony planet. They will be put ahead of their crew members, who have never been awake during the voyage.

"We do not necessarily have to produce anything new. The sleepers have not changed and are the same people. You can clone into them any beliefs you want before they are awakened. They will not be that hard to govern and control, especially since you are going to write the laws and have all the weapons. You can remove and eliminate the existing crew when we are ready if they prove too hard to control. You will need to have control of the robots to do this and a force of security officers."

So far, he sounds lucid, Sam thought.

"David, I will work on ideas for a cohesive plan that I think most people can get behind. It will be based on the old Earth religion. I, of course, want to be in charge of all things spiritual on this ship and the new home planet. I might have a plan for the continuation of some of old Earth's religious beliefs," said Sam.

"Sam, are you or Tom familiar with the Israelites' exodus from Egypt?" asked David. "This supposedly occurred 4,000 or maybe 5,000 years ago.

Both Sam and Tom answered no.

"This is the condensed version," said David. "This is exactly what we are doing now, is it not? It's God's will. The conditions on Earth and the chain of events we experienced are causing this migration to resemble the Israelites' experience exactly. We are all heading to an unknown promised land. This migration represents the will of God. With a little bit of cloning help, I can unite this crew and make them one flock for you. Just give me some time and resources in the form of church elders and enforcers; I guarantee it will happen."

David could see both men were taken back by this highly creative line of bullshit.

Sam and Tom nodded. "Yes, it just might succeed in the present surroundings," said Sam. "Start the new church secretly and keep them informed. You must stay in contact with us to receive any instructions we see fit to issue."

"Yes, I agree," said David. "I promise to get back to you through my father as soon as I develop the initial plan for the church cover-up and start-up."

David stood up with a great big smile on his face. It was as if it was clear to him now. He believed his life up to this point had been a test administered by God to see if he was worthy of this new beginning. David felt like he had finally found his calling, founding a new church for all the sinners on this ship. David was ecstatic as he exited the room.

They all wanted and hoped David's grandiose plan succeeded. Sam floated ideas for dealing with the Fleas.

"Dr. Harris was no longer a problem. Tom, we could expose the people in her department to the hypnotic suggestion and either kill them on the spot or use the mind trap if we get a reaction. We would have to use the robots; any of our human guards could be Fleas. Let us start the elimination process as soon as possible," Sam said. "Also, Tom, try to get the Colonel to increase the production of robots. I want the police model. Try and get me as many as you can."

CHAPTER 25

End Game

Not too long after the flip, the deep space radio receivers were brought back online. The radio room had become a lonely place, and there had not been any contact for over 500 years. The radio technicians had to be taken out of storage to get the equipment up and running. The ship also had Hubble-type telescopes and radio telescopes on board. They were all put back into working order.

They began the search for their new home, a planet named E-2 (Earth 2). The captain was sure that their navigation was not going to be perfect, and he expected to have to make minor adjustments. The sooner he made them, the easier they would be to perform.

After a number of years, the technicians reported to Tom the radio waves they were picking up could be explained. But there were still small oddities in the radio waves they could not identify. The search for E-2 was on schedule. The process of narrowing down its location was proceeding as expected.

The BC met again to discuss the last 130 or so years and the next one hundred years till they arrived at the new planet.

The captain announced that E-2 had recently been located. Everyone was ecstatic upon hearing the news. All the ship's resources and a sizable portion of the crew would be needed to prepare the ship's shuttles. Their thrust would provide the ship with only a small amount of maneuverability, but it would be enough. The most effective way for the ship to change direction in the dragless environment of space was to apply thrust 90 degrees to the ship's direction. The ship's main engine was stationary. Without the shuttles, the ship could not alter its course.

David turned out to be a touchy topic. Everyone expressed at least a small amount of skepticism about David's future.

"Major, what do you think your son's role in the future?" asked Sam.

"I don't know. I will abstain from the BC's decision, but I will back whatever course of action they decide on," he replied.

Sam could not believe his ears. The Major was putting his son's future in the hands of a roomful of psychopaths! Sam became even more suspicious of the Major. He still had nothing he could put his finger on. So far, David's new church remained a secret from the rest of the BC.

"I will manage David. He will not be a problem," said the Major.

This seemed to satisfy everyone.

"In effect, we have operational control of the ship," said Sam. "I see no reason to change that. Let us continue to eliminate Fleas and clone more security and police officers. We can, as officers become available, gradually station more of them throughout the ship. The Harris Fleas will not be a problem if we expose the crew to the hypnosis trigger on a regular basis."

"We accidentally discovered some of the s required as many as six separate exposures to the trigger phase before there was any reaction," interjected Tom. "The plan is to incorporate the Phase into our everyday life. Alice has been working on identifying the people in storage who were cloned in Harris's Lab. They will be dealt with at a later date. We are having remarkable success exposing the Harris Fleas."

Tom had some broad questions about cloning new people. "Cloning security and police officers was proceeding smoothly. Most crew members would just be everyday crew members performing mundane jobs every day. We have not yet decided what kind of personality and behavioral traits to give them," said Tom. "The easiest way would be to control their IQ, in other words, dumb down the crew. I propose having Alice research and decide the minimum IQ required to perform all the different job descriptions on the ship and we use that data to dictate what we create. We can also give them the proper physical characteristics and sex for each job. I have talked to Alice about the feasibility, and she assured me it would not be a problem to redesign people to all resemble one race. She only needed guidelines, and she would write the necessary codes for the cloning labs to use. Alice suggested using the Aryan race as the model.

"The BC should allow people to procreate the old-fashioned way through sex. Childrearing and sex will occupy most of the crew's time. Threats to their children will also keep them in line. This will, of course, eliminate our ability to control the child's characteristics, but after puberty, we can re-clone them and recreate them in their parents' mold. This, I think, will be an easy sell. After all, around 1,000 years ago, the same type of people gleefully destroyed their own unborn children. So, altering their children slightly should not present much of a problem for the parents."

Everyone then agreed removing one's desires and memories during cloning concerning same-sex relationships was a clever idea. It was believed this action would remove a lot of anxiety and hatred from society. Eventually this abnormal behavior would just fade away into the past, and society would just forget it ever happened.

Cloning was turning into a way to solve a lot of society's perceived problems without experiencing any pushback. Cloning was never forced on anyone; you lived and then died, and that was the end of you unless you agreed to be cloned into whatever the government wanted for you.

The BC decided to meet as needed or when someone called one. The suggestion was proposed by Sam and received unanimous agreement. Sam soon got the other co-conspirators together.

"How is the production of the robots coming along, Tom?" Sam asked.

"There will be approximately four hundred police robots under your control in the next couple of dozen years," he replied. "These robots will answer you and you only; for day-to-day service with their officer, you can set the robot's perimeters. The ability to control all the robots at once or separately will be available. You will also be able to see and hear everything the same as the robot in real time.

"I have also made some adjustments to the Colonel's robots. He is unaware of that. By the way, he also wanted 400 robots after he learned how many the police department was going to receive."

Sam did not like that one bit, and it showed.

Tom smiled. "Do not worry; I was not able to install the software necessary for you to take over his robots on the fly without his technicians' knowledge. I did the next best thing. You can remotely shut them all down. The sequence to restart them needs to be followed precisely, or they will be destroyed, and then it will take a major overhaul to get them up and running."

I am almost there. All I need is time, and I will be the last man standing. Sam thought.

His own police force was going to be the key to controlling the ship. He realized he needed a plan on how to eliminate his peers in the BC. The plan's only weakness was he had to trust Tom and the Major. But until he produced an alternative, he was locked in.

CHAPTER 26

A Big Surprise

The Captain called a BC meeting. He did not disclose the agenda. When they all gathered, he introduced the communications department head. He was a civilian, and no one except Sam knew who he was. But Sam did not reveal that fact. After introducing Dr. Silvers, he turned the meeting over to him.

Dr. Silvers spent the first 475 years in storage. He was only awakened when the commander who had been in charge of communications turned out to be a Flea. He had died in a shootout with the bridge crew and was dead before the mind trap could be employed. A records search revealed Dr. Silvers was the next most qualified person on board, so he was awakened as fast as possible. But he was still a person from Earth's 24th century; nothing about him changed, nothing at all. There had not been time. He was petrified by the people who surrounded him now. He easily recognized them for what they were.

"Dr. Silvers, you are in safe company," Sam said slowly and softly. "What has been discovered since you became the communications department head?"

"The communication techs had for some time now been aware of a strange radio signal," Dr. Silvers said. "It was not exactly a stand-alone radio signal."

There were blank looks everywhere.

Dr. Silvers tried again. "Think about a body of water and something is moving under the surface. But you can only see the surface ripples. So, you actually do not know what is under the water. Something has to be there to make the ripples. We, as yet, have no idea what to expect. The ripples are not random and appear to be created by something, but we do not know what.

"There is some good news. E-2 is not going to present too much of a problem for us to achieve orbit. The ripples appear to originate in the neighborhood of E-2. What this means, we do not know. I have requested that my department have available all the necessary resources to analyze and understand this anomaly. Now, I would like to be excused and return to my lab."

Dr. Silvers rose and walked out of the meeting. There was complete silence in the room. Then, suddenly, everyone began to talk at once. The Colonel pulled out his pistol and used it as a gavel to establish order.

Sam realized everyone had probably come to the same conclusion. What if the ripples originated from an alien source? If this was true, then there really was cause for alarm. But there was nothing anyone could do until they knew more.

The years continued to fly by. Then, unexpectedly, Dr. Silvers made a strange request. He wanted to wake up one of the world's foremost theoretical physicists to help with the ongoing research concerning the ripples. There was one caveat, do not make any changes except give him a younger body.

"I have had cloning since our meeting, and I have found it to be an immense help to occupy a younger body," Dr. Silvers said.

The BC was unanimous, and the cloning was accomplished as quickly as possible.

***

Sam, Tom, and the Major were steadily accomplishing all the preparations to take over the ship. Sam's justice department was up and running. The police force had steadily grown in numbers, and they now had officers patrolling every inch of the ship. Oliver had the drug business humming, and he had become an addict himself. Sam planned to leave him in place; he did not pose a threat to their plans in his present condition.

The Flea problem was not that hard to fix; it just took time. You could only clone a dozen or so at a time. Dr. Harris's department had proven the toughest nut to crack. Extra security guards had to be employed to maintain order numerous times. More than half of the Fleas were killed outright, with no chance to use the mind trap. This actually saved time. The replacement tech could be prepared and put to work, and they were not aware of what was happening around them when they arrived.

The plan to reduce the crew to the absolute minimum number was proceeding slowly. The workers in the drug business were already at the minimum number.

Sam had the police round up all the people who did not contribute to the operation of the ship, and they were placed in storage. This had proven to be a larger number than expected: 1,056 people, to be exact.

The replacement of the general crew was just getting started. The software programs to dumb down the crew were not available.

The techs had to do a great deal more research than planned to develop a program. Sam remembered as a child reading old history books about dumbing down the education system.

The few new crew replacements were already proving to be loyal and somewhat docile, exactly what he was looking for in a crew member. As the number of new model crew members increased, so did their children, and, for the first time ever, schools were being built. The BC had reluctantly given permission.

There was one strange anomaly. All the parents had a double-digit IQ, but their children's IQs spanned the full range of human intelligence. The children born from natural birthing with really high IQs would be exempt from the dumbing-down process. They would provide a valuable source of genetic material.

**

The captain called a BC meeting to discuss course changes.

"The shuttles would soon be aligned and ready to start the process of making the necessary changes. E-2's location had been confirmed," Tom announced. "Within 25 or so years, *New Hope* would be in orbit. We are getting close and now moving very slowly. We will try and match E-2's orbital speed around the sun. E-2 is the fifth planet in the solar system, containing fifteen planets, and it is number six in size. After arriving we now realize that possibly two other planets may be habitable. The course changes will not affect gravity, but as soon as we obtain orbit, the ship will go to almost zero gravity. We will issue magnetic shoes, the same as were used in the construction of the ship, to anyone who wants them."

"The work to get the ship ready for zero gravity had been underway for some time," explained the Captain. "The ship was going to return to its pre-launch condition. The work was almost completed. All the nonessential personnel would be landed on E-2 as soon as possible."

"All the calculations for the ships' last maneuver were already in progress as it slowly approached E-2's orbit," added Tom.

Sam and his coconspirators were ready. The only thing left to decide was when to act. Sam was beside himself; the planned takeover and elimination of the BC committee was at hand.

Suddenly, there was an urgent message. "BC members. This is the Captain. Drop whatever they are engaged in and report to the bridge immediately."

Dr. Silvers and the new physicist, Dr. Goldberg, were waiting. They both were white as cotton.

"Doctors, please continue explaining your findings," the Captain said. He turned on the ship-wide sound system so the entire crew would all hear the same message.

"We have with 100% certainty identified the ripples," stated Dr. Silvers. "I will spare you a long explanation. Dr. Goldberg and I both believe the ripples in the fabric of space are caused by faster-than-light radio waves or, as they are sometimes called, sub-space radio waves."

Everyone spoke all at once. "That is impossible!"

"Normally, we both would agree with you," replied Dr. Silvers. "Except, we can prove that sub-space radio waves exist. I will now play the message."

Until this is figured out, the mutiny would have to be postponed, thought Sam.

Then, Dr. Silvers turned on a speaker that was linked directly to a makeshift sub-space radio receiver.

"WE HAVE BEEN PATIENTLY WAITING FOR YOUR ARRIVAL."

Made in the USA
Coppell, TX
17 January 2026

67247899R00104